D0794094

# Government's Role in Innovation

# Government's Role in Innovation

By
**Dennis Patrick Leyden**
University of North Carolina
at Greensboro
**Albert N. Link**
University of North Carolina
at Greensboro

**Kluwer Academic Publishers**
Dordrecht/Boston/London

**Distributors for North America:**
Kluwer Academic Publishers
101 Philip Drive
Assinippi Park
Norwell, Massachusetts 02061 USA

**Distributors for all other countries:**
Kluwer Academic Publishers Group
Distribution Centre
Post Office Box 322
3300 AH Dordrecht, THE NETHERLANDS

**Library of Congress Cataloging-in-Publication Data**

Leyden, Dennis Patrick.
Government's role in innovation / by Dennis Patrick Leyden, Albert N. Link.
p. cm.
Includes bibliographical references and index.
ISBN 0-7923-9261-2
1. Technological innovations--Government policy--United States.
I. Link, Albert N. II. Title.
T173.8.L49 1992
338.97306--dc20 92-20984
CIP



*Printed on acid-free paper.*

Printed in the United States of America

*For Our Families*

# Table of Contents

# List of Figures

# List of Tables

# Acknowledgements

The writing of this book was made substantially easier because of the help of a number of individuals. We thank Zachary Rolnik, Senior Editor at Kluwer, for his support of the project; Melissa Stone for her long hours spent creating and revising the word-processing files that have become this book; Toni Fields for the preparation of the art work; Puneet Kapur for creating the index; and Sabrina Woodbery for general editorial support.

Finally, we wish to especially thank our wives, Peggy and Carol, for their moral support and understanding throughout the process of writing this book.

*Dennis Patrick Leyden*
Greensboro, North Carolina

*Albert N. Link*
Greensboro, North Carolina

# 1
# Defining Government's Role in Innovation

Innovation has long been recognized for providing benefits to society far beyond those that accrue to any particular participant in the private sector, and for being an important contributor to economic growth.[1] Unfortunately, in the absence of governmental intervention, private-sector participants will only consider their own particular benefits when choosing the appropriate level of commitment to the innovation process. This results in what economists call "market failure."

Market failure occurs whenever society's benefits and costs are not appropriately balanced. Market failure can arise for a number of reasons. Perhaps the best known examples are monopoly, where the firm has the ability to charge a price in excess of the cost of production, and pollution, where the costs imposed on third parties are not considered in the production and sale of the pollution-causing good.

In the case of research and development (R&D), and especially in the case of basic research, market failure is often a result of features intrinsic to the production of information. There is evidence to suggest that a market economy underinvests in the production of knowledge because the firm that produces the knowledge is unable to capture fully all of the profits that arise from its creation.[2] Knowledge is created through innovative processes, and, as will be discussed below, R&D expenditures are a critical input into those innovative processes. As a result, the inability to appropriate fully the benefits accruing from the creation of knowledge leads to an investment in R&D that is, from a social perspective, too small.

The cost of inadequate investments in innovation is particularly high in today's globally-competitive environment where continued technological advancements are critical to sustaining, if not advancing, the economic prosperity of the United States. Many indicators point to the fact that American industry no longer competes well. U.S. companies' share of the domestic American market in technologies pioneered domestically has been declining for two decades.[3] In fact, as evidenced in the U.S. Department of Commerce report on emerging technologies, the United States is no longer the uncontested world leader in most emerging technologies.[4] This trend is especially noticeable in the U.S. semiconductor industry. In 1970, Japan's share of the world market in dynamic random access memories (DRAMs) was zero. It rose to 4.2 percent in 1975, 39.4 percent in 1980, and is now over 80 percent.[5]

While it cannot solve these problems singlehandedly, the government does have a crucial role in ensuring that society's general interests in innovation are represented in private-sector decision making through a variety of programs and initiatives that reward innovation at all levels in a way that creates as few countervailing distortions as possible. The broad spectrum of activities associated with fulfilling this role fall into two general categories: (1) the creation and maintenance of a legal environment conducive to private-sector investment in innovative activities, and (2) the provision of sufficient stimuli to overcome the natural inclination of private parties to consider only their private benefits when choosing the level of innovative activity in which to engage. Patents and the relaxation of antitrust activity are the primary means by which the government creates a legal environment conducive to private-sector innovative activity, while the provision of stimuli takes a variety of forms ranging from governmental grants and contracts to targeted tax incentives.

Interestingly, American voters seem to recognize the importance of government's role. In a recent poll taken by the Council on Competitiveness, American voters by a strong majority expressed a desire for the Federal government to take an active role in restoring

America's economic standing in the world market. When asked, "Which of the following statements comes closer to your point of view?"

> *We should rely on the private sector and free enterprise system to promote America's economic competitiveness and ability to deal with foreign competition, and government should not get directly involved;*
>
> *The Federal government should play a direct and active role in working with business to promote America's economic competitiveness and ability to deal with foreign competition;*

sixty-one percent chose the second statement.[6]

Perhaps reflecting that consensus, the Bush Administration, in what many view as this country's first official statement on technology policy, set forth the view that:[7]

> ... A nation's technology policy is based on the broad principles that govern the allocation of its technological resources. Competitive market forces determine, for the most part, an optimal allocation of U.S. technological resources. Government can nonetheless play an important role by supplementing and complementing those forces.... The principal role of the Federal Government will be to provide an environment conducive to long-term economic vitality, and not allow special interests to divert attention or resources from this goal.

Although this policy is still in the formative stage, a number of areas of Federal responsibility were outlined in the technology policy statement. Of particular note were the following recommendations:

- Increase Federal investment in support of basic research (especially at universities).

- Participate with the private sector in precompetitive research on generic, enabling technologies that have the potential to contribute to a broad range of government and commercial applications.

- Continue the Federal government's development of products and processes for which it is the sole or major consumer.

- Streamline Federal decisionmaking structures and mechanisms to eliminate unnecessary and cumbersome regulations and practices that inhibit industrial competitiveness.

- Encourage international cooperation in science and technology where mutually beneficial, and inform U.S. researchers of opportunities to participate in R&D initiatives outside the United States.

- Improve the transfer of Federal laboratories' R&D results to the private sector.

- Promote increased collaboration between industry, Federal laboratories, and universities.

- Expedite the diffusion of the results of Federally-conducted R&D to industry.

Although somewhat vacuous in terms of the specifics for an implementation strategy, this policy statement does highlight a number of important areas in which the government acknowledges its role in innovation. Perhaps most importantly, this policy statement acknowledges the role of the Federal government in three key areas: (1) funding of R&D performed in the private sector, (2) funding of Federal laboratory research activities and the effective transfer of that knowledge to the private sector, and (3) funding of basic research (especially at universities) and encouraging of industry-university research relationships. It is these three areas of research, more than any other, that generate technologies fundamental to increasing the rate of technological development in the private sector, and it is these three areas that provide the subject matter for this book.

To understand more fully the importance of Federal funding of private R&D, Federal laboratory activity, and the funding of basic

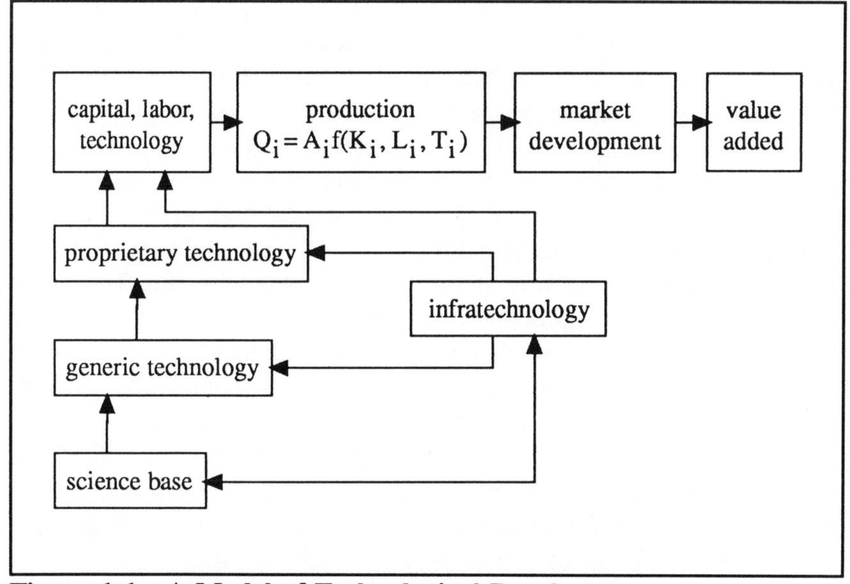


**Figure 1.1. A Model of Technological Development**

research, consider the model of technological development illustrated in Figure 1.1.[8] Embedded in this model is a simple linear view of economic activity in which factors of production -- capital, labor, and technology -- enter the production process, and an output results. Market conditions then give the product economic value. Supporting the technology input are several key technology elements: proprietary technology, generic technology, infratechnology, and the science base.

The science base provides the foundation for much of the new technology seen today. This base comes from basic research that is funded primarily by the public sector. In 1989, nearly 60 percent of the $14.7 billion spent on university-based basic research was Federally-financed.[9] Basic research is the search for fundamental scientific principles without consideration of practical applications.[10] Discovery of these principles does not necessarily lead, nor is it intended to lead, to a new product or process application. Thus, because the science base does not consist of appropriable knowledge, incentives are lacking for full support by the private sector. About 50 percent of all basic research is conducted at universities, although some is conducted directly by governmental agencies when it is germane to their mission.

Infratechnologies are a less widely recognized element of an industry's technology base. As the name implies, they are technologies that facilitate R&D, production, and marketing in industries. Infratechnologies include evaluated scientific data used in the conduct of R&D; measurement and test methods used in research, production control, and acceptance testing for market transactions; and various technical procedures such as those used in the calibration of equipment. Infratechnologies are not generally embodied in an industry's product technology in the same way as the generic technology (the knowledge that is organized into the conceptual form of an eventual application). Instead, infratechnologies facilitate the development of the generic technology by providing highly precise measurements and creating organized and evaluated scientific and engineering data necessary for understanding, characterizing, and interpreting relevant research findings. Typically, they tie at root to the fundamental units of measurement. They also provide the measurement and testing concepts and techniques that enable higher quality and greater reliability at a lower cost of production. Finally, infratechnologies provide buyers and sellers with mutually acceptable, low-cost methods of assuring that specific performance levels are met when technologically-sophisticated products enter the marketplace.

Clearly, then, the science base and related infratechnologies lie in the public domain. They represent knowledge that is nonproprietary, broad in scope, and widely available for use. Accordingly, they are financed primarily by the public sector, though more private support is provided for infratechnology than for basic research.[11]

Generic technology and the associated research process represents the organization of knowledge into the conceptual form of an eventual application and the laboratory testing of the concept. Its foundation is usually the science base, but unlike scientific knowledge it has a functional focus. Generic technology tends to have mixed proprietary and nonproprietary characteristics; that is, it is quasi-public with the degree of publicness based on the stage of maturity of the industry and the maturity of the technology. For example, the basic design concepts and architecture of integrated circuits are a generic technology. Originally, integrated circuit architecture and design concepts were proprietary, coming from privately-financed applied research. But today, the same concepts are in the public domain in the sense that they are incorporated in all integrated circuit designs. Because of the quasi-public nature of the generic technology, firms, and often the government, underinvest in it.

Proprietary technologies are in the private domain; they are fully appropriable by the firm so long as they are secret. Such product and process technologies result from self-financed R&D or, more precisely, from development activities concerned with creating marketable technology-based products or processes.

These elements of industrial technology are interrelated as the arrows in the figure suggest. Competitive survival in any advanced economy such as ours will depend upon technology-based strategies that emphasize the integration of internal R&D (proprietary technology) and external sources of technology (the science base and infratechnology) into one focused development process. As this occurs, and it must occur to meet the new order of competition in the world, the role of government in innovation will become more critical.

The chapters that follow examine in detail past and expected future Federal governmental involvement in financing private-sector R&D activities, Federal laboratory research activities designed to support private industry, and basic and generic research conducted in universities and state-based science centers. Through a detailed examination of government's past and future efforts at fulfilling its role in innovation, our intent is to reveal the essential qualities of governmental involvement that stimulate both innovation and productivity growth.

The examination begins in Chapter 2 with a brief summary of the academic literature related to the relationship between Federally-financed R&D and private sector innovative activities, and between Federally-financed R&D and productivity growth. A large body of evidence from research conducted over a considerable period of time shows that governmental R&D allocations do stimulate private R&D investments. Additional evidence suggests that governmental R&D allocations also stimulate productivity growth. See Figure 1.2. Interestingly, though the policy implication of these conclusions would appear to be clear, a closer inspection of the issue reveals that much more needs to be known before a rational policy for stimulating innovation can be developed. As Figure 1.1 revealed, R&D is not a black box; it is instead a complex process beginning with activities that are in large part not appropriable, and finishing with activities that are fully appropriable. Does it make a difference whether governmental efforts intended to stimulate innovation are directed at the private R&D process generally or at specific components of the private R&D process? As discussed at the beginning of this chapter, theory would suggest that government's role should be

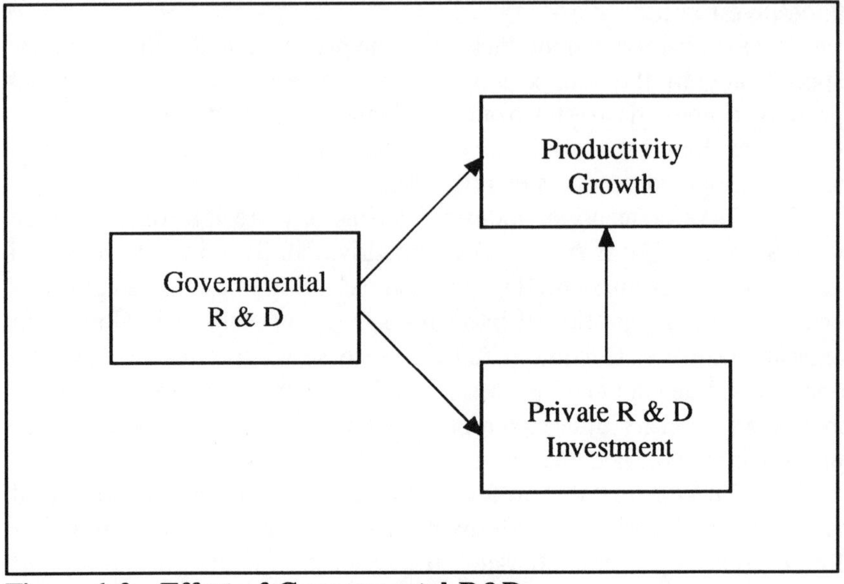


**Figure 1.2. Effect of Governmental R&D**

proportional to the degree to which private firms cannot appropriate the benefits associated with their efforts. The studies reviewed in Chapter 2 have little to say about this important policy issue.

Chapter 3 provides some insight into this issue. Using a theoretical framework to examine the connection between Federally-financed R&D and private R&D investment and supported by empirical evidence, Chapter 3 concludes that Federally-financed R&D stimulates private R&D investment because it contributes to the expansion of the private-sector firm's infratechnology base. Infratechnology is the critical factor because infratechnology essentially leverages the value of private R&D investment to the firm, thus compensating for the lack of appropriability typical in infratechnology investment. Because the sharing of knowledge with other firms in the industry is an important part of increasing a firm's infratechnology base, Federally-financed R&D also stimulates the sharing of knowledge across firms.

Of course, the question remains whether such stimulation of private R&D investment has any effect on productivity growth. Chapter 4 answers that question by surveying the level of infratechnology investment in both Federal laboratories and industrial manufacturing

R&D laboratories, and by examining the effect of such investment on productivity growth. Though not a comprehensive study of all innovative activity, the conclusion is clear: Federally-financed R&D increases productivity growth because it stimulates the private-sector's infratechnology base.

Government also attempts to stimulate innovation through the support of research activities undertaken in Federal laboratories. Chapter 5, through case studies, investigates the industrial and social benefits emanating from Federal laboratory investments in two research areas: optical fiber and electromigration characterization. Because these investments are primarily investments in infratechnology, the conclusions are similar to those found for Federally-financed R&D: Federal laboratory research increases the efficiency of private R&D activity and stimulates growth.

The last area of governmental involvement in directly stimulating innovative activity is the Federal funding of basic and generic research at university and state-based science centers. Chapter 6 examines the influence of university and state-based science center research on firm performance. Of particular interest is the degree to which firms rely on these external sources of technical knowledge, and the extent to which this reliance affects their performance. Much of the research conducted in such facilities is Federally-financed, and the nature of the research is primarily basic and generic. Again, because basic and generic research is typically nonappropriable, investment in such research has benefits similar to investment in infratechnology: private R&D and productivity growth are stimulated.

While Chapters 2 through 6 investigate direct governmental involvement in various R&D processes, Chapter 7 examines in detail an indirect policy initiative that has the potential to affect, to a significant degree, innovative activity in the private sector: the R&D tax credit. Because the R&D tax credit, unlike the measures discussed in the previous chapters, is an indirect method of stimulating innovative activity, it tends to provide general stimulation of all categories of private R&D activity, including those activities that are for the most part appropriable. While this too provides social benefits, it would appear much less effective than those efforts that target those types of R&D that are less appropriable, namely, infratechnology, basic research, and generic research.

Finally, Chapter 8 summarizes the general themes in the previous chapters, evaluates the efficiency with which the Federal government has fulfilled its role in innovation, and proposes new ways for the Federal government to fulfill its role more effectively in the future. Of particular interest is the newly formed Advanced Technology Program which offers the hope of stimulating those types of innovative activities that theory and empirical evidence suggest are most valuable.

## NOTES

1. Cohen and Noll (1991), in their book on Federal R&D projects to stimulate private-sector commercial technology, provide a review of the empirical evidence of the link between innovation and economic growth. They also discuss other "quasi-economic" benefits of innovation including national prestige, national security (both directly through defense-related technologies and indirectly by providing the Nation with leverage in international relations), and political stability.

2. Bozeman and Link (1983), for example, summarize these arguments.

3. See Council on Competitiveness (1989).

4. See U.S. Department of Commerce (1990).

5. See Carnegie Commission (1991) and Council on Competitiveness (1989).

6. See Council on Competitiveness (1991). In fact, this poll also revealed that 73 percent of these voters blame the U.S. government "a great deal" or "quite a bit" for the competitive problems now facing this country.

7. See Executive Office of the President (1990).

8. This model was proposed in Link and Tassey (1987) and recently extended in Tassey (1991, 1992).

9. See National Science Board (1991) for these data -- the most recent data that are available.

10. The National Science Foundation (1990) definition of basic research includes "the cost of research projects which represent original investigation for the advancement of scientific knowledge and which do not have specific immediate commercial objectives (although they may be in the fields of present or potential interest to the ... company)."

11. According to Link (1991), a survey of 62 Federal laboratories indicated that 29 were engaged in 1990 in infratechnology research. In that year these 29 laboratories invested $1.2 billion in infratechnology research, or, on average, 38 percent of their total budget. See Chapter 4.

# 2
# The Influence of Federally-Financed R&D

Though a variety of reasons are often given for supporting Federally-financed R&D, the most common justifications have been that it stimulates private-sector innovative activity and that it increases the Nation's rate of productivity growth.[1]

After briefly describing the history of Federal support for innovative activity, this chapter summarizes attempts by researchers to measure the effect of Federally-financed R&D on private R&D investment and on productivity growth.

## HISTORY OF FEDERAL R&D SUPPORT

Historically, public policies toward innovation have been directed at R&D activity, and the government has a long history of such support. The U.S. Navy's sponsored research programs can be traced as far back as 1789, and the U.S. Department of Agriculture's involvement in the land-grant college system dates from the mid-1800s.

Since World War II, direct governmental support of R&D has

increased dramatically in response to military needs and to the government's responsibility for enhancing research capabilities as outlined in the National Science Foundation Act of 1947. This support has been focused in two areas. One is basic research, which adds to the Nation's (and, in fact, the world's) science base. The other is applied R&D which, even when defense-oriented, enhances the research capabilities of individual firms.

Total direct Federal outlays for R&D rose steadily from World War II to the mid-1960s. Federally-financed R&D as a percentage of the total Federal budget increased from 1.7 percent in 1945 to a high of 12.6 percent in 1965. After 1965, the percentage decreased continuously, reaching a low of 5.2 percent in 1981. It was 6.1 percent in 1990.[2]

While the Federal government is a major source of R&D funding, most R&D is performed in the private sector, in particular by firms in manufacturing industries. In 1990, the Federal government funded nearly 44 percent of all R&D, but performed only 11 percent.

Direct governmental support of university research began with the land-grant college system. Over time, that support has grown so that today governmental support for general research at universities and colleges is an important element of the Nation's science base. However, in recent years governmental support to the university system has been slipping. Total Federal obligations to universities in constant dollars rose at an average annual rate of 5.5 percent between 1963 and 1975, and rose at an average annual rate of 4.8 percent between 1975 and 1980. From 1980 to 1981, total funding actually fell by 15.3 percent. While funding rose between 1983 and 1985, it is now falling again.

Indirect governmental support for industrial R&D is tied directly to the adoption of section 174 of the 1954 Internal Revenue Code, which codified and expanded tax laws pertaining to firms' R&D expenditures. While direct financial support to industry has been cyclical, indirect financial support has increased over time as evidenced by a number of recent policy initiatives such as the R&D tax credit portion of the Economic Recovery Tax Act of 1981, and its recent renewals;[3] the National Cooperative Research Act of 1984;[4] the Technology Transfer Act of 1986; and the Omnibus Trade and Competitiveness Act of 1988 which established the Advanced Technology Program.[5]

## THE EFFECT ON PRIVATE R&D

Early on, researchers were interested in the effect of Federal R&D funding on the private sector. Spurred by the work of Blank and Stigler (1957), empirical studies were undertaken to investigate how governmental R&D influences the level of private R&D investment. With minor exceptions, the empirical evidence supports there being a complementary relationship between private R&D and governmental R&D.[6]

For the most part, these empirical studies have employed cross-sectional data (firm-level or industry-level data) to estimate a simple, single-equation model of the form:

$$F = g(G,\mathbf{X}) \tag{2.1}$$

where F is the level of private funds allocated by the firm (or industry) to R&D, G is the level of governmental financial support for R&D received by the firm (or industry), and **X** is a vector of other factors included as control variables that reflect such factors as firm size, industry concentration, and product-line diversification.

Typically, empirical studies using models like the one in equation (2.1) find $\partial F/\partial G > 0$. This is interpreted to mean that governmental R&D complements private R&D. Surprisingly absent from all of these empirical investigations is a theoretical foundation that characterizes the mechanisms through which governmental R&D and private R&D interact. Such a broader understanding of the role of governmental R&D is imperative if one is to begin to evaluate fully the returns to direct governmental support of industrial innovation and to design rationally programs to support future R&D activities. Chapter 3 is devoted to developing such a foundation.

## THE EFFECT ON PRODUCTIVITY GROWTH

Equally important as an understanding of the mechanisms through which governmental R&D and private R&D interact is an understanding of the mechanisms through which governmental R&D affects productivity growth. At this juncture, however, the academic literature is still grappling with basic issues and has not established any robust empirical findings.

Beginning in the early 1960s, researchers began to investigate quantitatively the impact of R&D spending (in total) on productivity growth. Typically assumed is a three-factor production function such as:

$$Q_i = A_i\, f(K_i, L_i, T_i) \tag{2.2}$$

where $Q_i$ represents output of the ith industry (or industrial sector), $A_i$ is an industry- or sector-specific Solow-neutral disembodied shift factor, $K_i$ and $L_i$ represent the industry's (or sector's) stock of physical capital and human capital (labor), and $T_i$ is the industry's (or sector's) stock of technical capital.

If the production function $f(\cdot)$ is Cobb-Douglas, then equation (2.2), after dropping the subscripts, becomes:

$$Q = A\, e^{\lambda t} K^{\alpha} L^{(1-\alpha)} T^{\beta} \tag{2.3}$$

where $\lambda$ is a disembodied rate-of-growth parameter and $\alpha$ and $\beta$ are output elasticities. Constant returns to scale are assumed only with respect to K and L.

Using logarithmic transformations and differentiating equation (2.3) with respect to t:

$$Q'/Q = \lambda + \alpha\, (K'/K) + (1-\alpha)\, (L'/L). \tag{2.4}$$

From equation (2.4):

$$A'/A = \lambda + \beta\, (T'/T). \tag{2.5}$$

The parameter $\beta$ in equation (2.5) is the output elasticity of technical capital from equation (2.3):

$$\beta = (\partial Q/\partial T)(T/Q). \tag{2.6}$$

Substituting the right-hand portion of equation (2.6) into equation (2.5):

$$A'/A = \lambda + \rho\, (T'/Q) \tag{2.7}$$

where $\rho = (\partial Q/\partial T)$ is the marginal product of technical capital and T' is the change in units of net private investment into the stock of technical capital.

Empirically, if it is assumed that the stock of R&D-based technical capital does not depreciate, or depreciates very slowly, then T′ can be approximated by the flow of R&D expenditures in a given time period, RD. Hence, equation (2.7) gives rise to the empirical model:

$$(2.8) \qquad A'/A = \lambda + \rho\,(RD/Q) + \epsilon.$$

Estimates of $\rho$ from equation (2.8) consistently confirm the productivity growth-enhancing value of R&D.[7]

A few researchers have estimated an alternative version of equation (2.8), dividing the regressor into two parts:

$$(2.9) \qquad RD/Q = RD^{P}/Q + RD^{G}/Q$$

where the superscript P refers to privately-financed R&D, and the superscript G refers to the Federally-financed R&D.[8] Their findings consistently suggest that governmental R&D has an independent, positive impact on productivity growth.

Clearly, therefore, governmental support of private-sector R&D activities provides an important social benefit by increasing productivity growth. Unfortunately, researchers to date, present company included, have pursued an empirical avenue of interest in the absence of a complete understanding of the underlying mechanisms through which governmental R&D affects the production process. Thus, only the most general policy implications can be drawn from such a result.

# NOTES

1. The three classic roles for government are to enhance efficiency, promote equity, and engage in stabilization policies. The justification for stimulating private-sector innovation *per se* can be debated. It seems clear, however, that government does have a role in stimulating growth as part of a comprehensive stabilization policy. See Musgrave and Musgrave (1989).

2. Unpublished National Science Foundation data. Other historical statistics in this chapter came from various National Science Foundation publications.

3. See Chapter 7.

4. See Link and Bauer (1989).

5. See Chapter 8.

6. See Carmichael (1981); Levin and Reiss (1984); Levy (1990); Levy and Terleckyj (1983); Leyden, Link, and Bozeman (1989); Lichtenberg (1984, 1987); Link (1982, 1987); Mansfield and Switzer (1984); and Scott (1984).

7. See reviews by Griliches (1988) and Link (1987).

8. See Link (1981) and Terleckyj (1974).

# 3
# The Production of Technical Knowledge

For decades, researchers have investigated the government's general role in stimulating private-sector innovation. However, with the productivity-growth slowdown in U.S. industries that began in the mid-1960s and that accelerated in the early 1970s, researchers began to focus on the direct impact of governmental actions on private-sector research and development (R&D), asking particularly whether governmental R&D allocations to industry complement or substitute for privately-financed R&D. While empirical evidence supported a complementary relationship, absent from all investigations was a theoretical framework to explain why such a complementary relationship exists.

Clearly, a broader understanding of the mechanism through which governmental R&D and private R&D interact is imperative if one is to begin to understand the subtleties of the existing empirical findings and, more importantly, to evaluate fully the returns to direct governmental support of industrial innovation. As evidenced by provisions in the Stevenson-Wydler Act, the Bayh-Dole Act, and the Technology Transfer Act, policy makers are challenging, more than ever, the view that public-domain research provides the greater social benefit. To address such an issue intelligently, more must be known about the benefits from conventional governmental funding to industry and the ways that governmental R&D and private R&D interact.

This chapter argues that infratechnology provides the critical link between governmental R&D allocations and private R&D funding; and that the observed complementarity between these two types of funding is itself the result of technical complementarity at the production level between funding, infratechnology, and knowledge sharing.[1] Single-equation analyses performed heretofore have generally ignored the role of infratechnology and knowledge sharing. As a result, they have left the impression that the observed complementarity between governmental R&D and private R&D funding is a predetermined outcome of the R&D process. We argue that the observed complementarity is not predetermined. Rather, it is the result of a complex balancing of forces whose origins are found in standard production convexities and in the technical complementarity referred to above.

The first half of this chapter contains a theoretical formalization of these ideas.[2] We examine a profit-maximizing firm engaged in a private R&D process in which both basic research, and applied research and development activities are conducted. In the absence of a governmental R&D allocation, the firm maximizes profits by choosing a level of private R&D funding and dividing that funding in an optimal way between basic research, and applied research and development.

Critical to the firm's productivity is the level of infratechnology available to it. Infratechnology, as discussed in Chapter 1, is used to facilitate the R&D process and may be embodied either in particular individuals or in such things as structures used to house R&D activities; equipment used to conduct R&D activities; or pre-existing knowledge used to understand, characterize, or interpret the R&D process.[3] We assume that the infratechnology in an industry is generally available to all firms, but that its effective use depends both on the firm's level of R&D expenditure and on the sharing of knowledge in a *quid pro quo* manner.[4] While the sharing process may impinge on the firm's competitive position, it nevertheless provides the firm with the potential to increase its infratechnology and thus enhance its entire R&D process.

Infratechnology also plays a critical role if the firm receives a governmental R&D allocation. For the government, an R&D allocation provides a means for it to exploit benefits inaccessible otherwise. While the identification of these benefits will depend upon the specific project, in general these benefits center around the ability of the government to lower the cost of conducting a given R&D project and the ability of the government to shift either production or political risks to those willing to bear the risk (at a price). For the firm, the receipt of a governmental R&D allocation allows it to conduct an R&D process alongside its private R&D process. Because both the governmental R&D process and the private R&D process are housed within the same firm, both will contribute to and share the same infratechnology. Thus, the receipt of a governmental R&D allocation increases the firm's level of infratechnology and hence its productivity and profitability in the private sector.

The second half of this chapter reports the results of an empirical investigation based on the theoretical framework developed in the first half of the chapter. Because of data limitations, a simplified model derived from the theoretical framework is developed and evaluated using a unique data set drawn from U.S. industrial R&D laboratories. In both an initial descriptive analysis as well as in a more formal regression analysis, our results confirm the findings of others of the existence of complementarity between governmental R&D and private R&D. Beyond that, however, we find evidence consistent with the hypothesis that the complementarity between governmental R&D and private R&D is the result of technical complementarity in production between funding, infratechnology, and knowledge sharing.

Finally, it should be noted that for the sake of clarity as well as because of data limitations, we have ignored important issues of uncertainty and dynamic process. Though we believe that the effects of uncertainty can be best understood by first understanding outcomes generated by a deterministic framework, the problem of uncertainty is sufficiently important that it should be recognized, if only in a tangential way. The appendix to this chapter contains a model that explores the effect of uncertainty in the particular context of a risk-averse, budget-maximizing bureaucracy.

## THEORETICAL FRAMEWORK

### R&D Production Processes

Consider a single firm that engages in private R&D in order to increase its profits, and assume that this firm receives a governmental

R&D allocation (e.g., a contract) in return for engaging in a separate R&D process of interest to the government. The private R&D process includes both basic research as well as applied research and development. Basic research culminates in basic knowledge used in the applied research and development process. The output of the applied research and development process is applied knowledge used by the private-sector firm in the production of goods and services sold in private markets. The governmental R&D culminates in governmental technological knowledge used by the government in the production of goods and services desired by the government.

The degree to which the outputs of the private R&D process and the governmental R&D process are similar will vary with the particular nature of the firm's market interest and the type of goods and services the government requires. It is likely, however, that the output of the private R&D process and the output of the governmental R&D process will be neither homogeneous nor fungible. For simplicity, we assume that the two research processes are distinct, and that each is characterized by a separate production function. The existence of separate production functions, however, need not imply the absence of linkages between the two processes. Even if the production functions for private R&D and governmental R&D are separate, it is reasonable to assume that they will share the same infratechnology if conducted within the same firm.

**Private R&D.** The private R&D process is a two-step process beginning with basic research and ending with applied research and development activities. Define $\beta$ to be the relative amount of basic knowledge produced from the basic research process, and assume that the production function for $\beta$ can be written as a positive, strictly-concave function of the funds, $F_\beta$, devoted to that process and the relative level of infratechnology, I, available to the firm:[5]

$$\beta = \beta(F_\beta, I). \tag{3.1}$$

$\beta(\cdot)$, given the assumptions on the determination of basic knowledge, will have the following derivative signs:

$$\partial\beta/\partial F_\beta > 0$$

$$\partial\beta/\partial I > 0$$

$$\partial^2\beta/\partial F_\beta^2 < 0$$

$$\partial^2\beta/\partial I^2 < 0.$$

We assume in addition that the level of $\beta$ is nonnegative and that some injection of funds is necessary to produce a positive amount of basic knowledge:

$$\beta(F_\beta,I) \geq 0 \;\; \forall \;\; F_\beta,\beta$$

$$\beta(F_\beta,I) = 0 \;\text{ if }\; F_\beta = 0.$$

Figure 3.1 provides a schematic representation of the determination of $\beta$.

Finally, note that the sign of the cross-partial derivative $\partial^2\beta/\partial F_\beta \partial I$ was not specified above. We make no assumption regarding the ability of infratechnology to increase the productivity of the funds used to conduct basic research. This issue, of course, lies at the heart of the value of infratechnology and is examined in detail later in this chapter.

Let $\alpha$ be the relative amount of applied knowledge produced by the applied research and development process, and assume that the production function for $\alpha$ can be written as a positive, strictly-concave function of the funds, $F_\alpha$, devoted to the applied research and development process, and the relative level of basic knowledge, $\beta$, created through the process noted by equation (3.1):[6]

$$\alpha = \alpha(F_\alpha,\beta). \tag{3.2}$$

Thus, $\alpha(\cdot)$ has the following derivative signs:

$$\partial\alpha/\partial F_\alpha > 0$$

$$\partial\alpha/\partial\beta > 0$$

$$\partial^2\alpha/\partial F_\alpha^2 < 0$$

$$\partial^2\alpha/\partial\beta^2 < 0.$$

We assume additionally that $\alpha$ is nonnegative, that some injection of

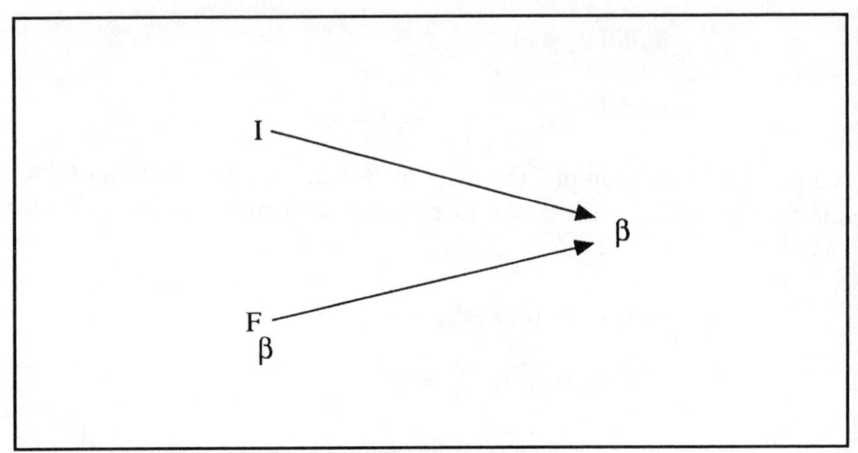


**Figure 3.1. Production of Basic Knowledge**

funds is necessary to produce a positive amount of $\alpha$:

$$\alpha(F_\alpha,\beta) \geq 0 \quad \forall \; F_\alpha,\beta$$

$$\alpha(F_\alpha,\beta) = 0 \quad \text{if } F_\alpha = 0,$$

and we make no assumption concerning the sign of the cross-partial derivative, $\partial^2\alpha/\partial\beta\partial F_\alpha$. A schema representing the applied research and development production process is provided in Figure 3.2.

**Governmental R&D.** Define $\gamma$ to be the amount of governmental technological knowledge produced by the governmental R&D process, and assume that the production function for $\gamma$ can be written as a positive, strictly-concave function of both the funds, G, devoted to the governmental R&D process and the relative level of infratechnology, I, available to the private-sector firm:

(3.3) $$\gamma = \gamma(G,I).$$

$\gamma(\cdot)$ will therefore have the following derivative signs:

$$\partial\gamma/\partial G > 0$$

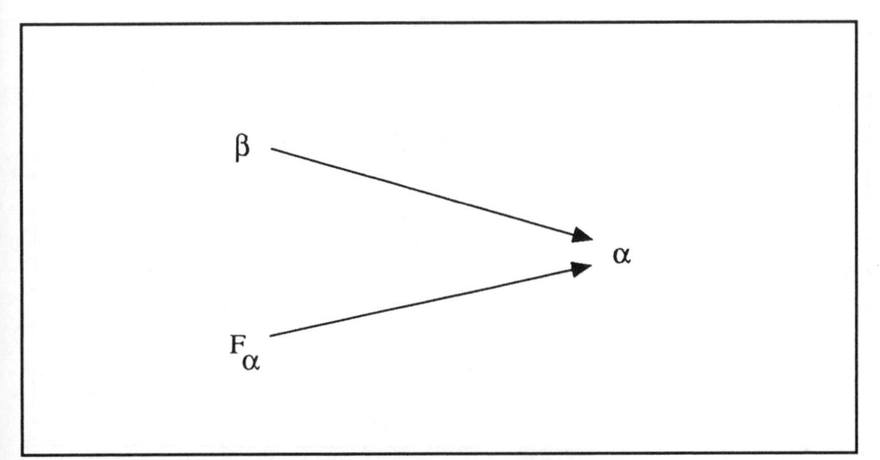

**Figure 3.2. Production of Applied Knowledge**

$$\partial\gamma/\partial I > 0$$

$$\partial^2\gamma/\partial G^2 < 0$$

$$\partial^2\gamma/\partial I^2 < 0.$$

As with the production of $\beta$ and $\alpha$, we assume that $\gamma$ is nonnegative and that some injection of funding is required to produce a positive $\gamma$:

$$\gamma(G,I) \geq 0 \quad \forall \; G,I$$

$$\gamma(G,I) = 0 \quad \text{if } G = 0.$$

Figure 3.3 represents a schema of the governmental R&D production process.

**Infratechnology.** The relative amount of infratechnology is assumed to be a positive, strictly-concave function of the total amount of funds devoted to the various R&D processes in which the firm engages, the level, S, of the firm's activity in sharing intellectual activities (such as conferences and symposia), and the level, M, of R&D activity conducted by the firm's competitors. Defining F to be the aggregate amount of funds devoted by a private-sector firm to its private R&D processes:

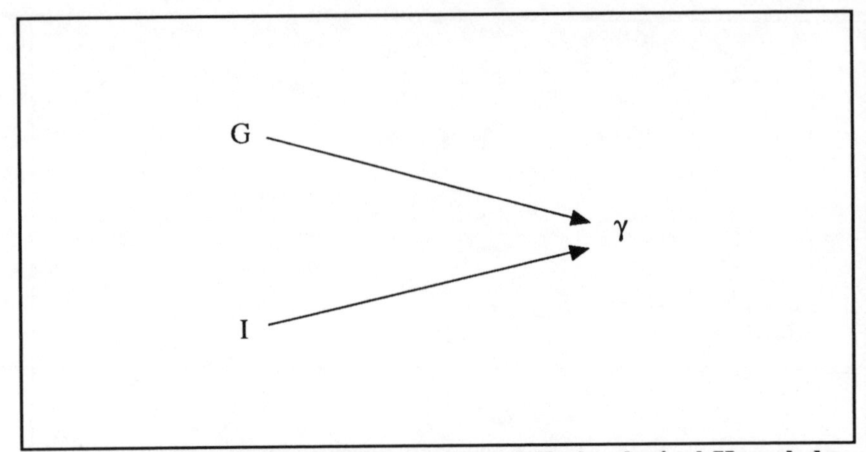


**Figure 3.3. Production of Governmental Technological Knowledge**

(3.4) $$F = F_\alpha + F_\beta,$$

the production of infratechnology can be described by the equation:

(3.5) $$I = I(F+G,S,M).$$

Total R&D funding has a positive but diminishing role in the production of infratechnology:

$$\partial I/\partial(F+G) > 0$$

$$\partial^2 I/\partial(F+G)^2 < 0.$$

S and M represent in a simple way the importance of spillovers to the R&D process. As has been noted by others,[7] the acquisition of knowledge is not free and typically involves large investments. While knowledge can sometimes be acquired in a one-way transfer, the acquisition of knowledge often involves a *quid pro quo* process in which the firm must share its knowledge with the outside world if it is to acquire outside knowledge. We assume that the relative level of useful infratechnology available to the private-sector firm is a positive function of S for low levels of S, but that it eventually becomes negative beyond some critical level of sharing $S_o$:

$$\partial I/\partial S > 0 \text{ for } S < S_o$$
$$\partial I/\partial S < 0 \text{ for } S > S_o$$

$$\partial^2 I/\partial S^2 < 0.$$

See Figure 3.4 for a graphical illustration of this relationship.

The effect of M on the firm's relative level of infratechnology will depend on whether cooperative sharing arrangements exist. If the firm does not engage in such arrangements, a rise in the R&D activity of the industry will reduce the relative value of the firm's infratechnology as other firms invest more in their own infratechnology. However, if the firm engages in cooperative sharing arrangements, it may be able to benefit from increased R&D activity elsewhere in the industry and thus mitigate, to some degree, the negative effect of the rise in M noted above. Thus:

$$\partial I/\partial M_{\text{no sharing arrangement}} < 0$$

$$\partial I/\partial M_{\text{no sharing arrangement}} < \partial I/\partial M_{\text{sharing arrangement}}.$$

Whether firms actually improve the relative level of infratechnology as a result of participating in sharing arrangements (that is, whether the derivative $\partial I/\partial M_{\text{sharing arrangement}}$ is positive) is an empirical question.

Private-sector firms engaging in both basic research, and applied research and development typically make funding decisions by first choosing the total level of private R&D spending and then dividing that sum between basic research, and applied research and development. Let $\rho$ represent the proportion of F going to basic research. Thus:

$$F_\beta = \rho F \qquad (3.6)$$

$$F_\alpha = (1-\rho)F \qquad (3.7)$$

where $\rho$ is restricted to the closed positive interval:

$$0 \leq \rho \leq 1. \qquad (3.8)$$

Hence, combining equations (3.1), (3.2), (3.3), and (3.5), the entire private R&D production process for the firm engaged in both basic research, and applied research and development can be described by the

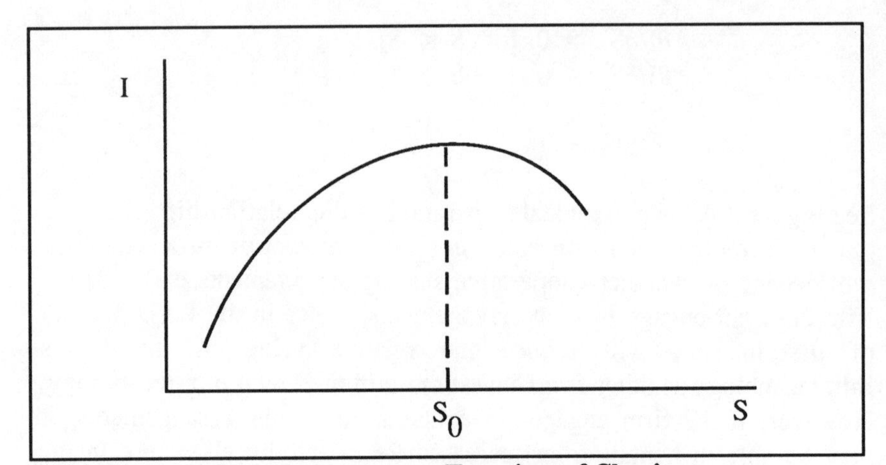


**Figure 3.4. Infratechnology as a Function of Sharing**

function:

$$(3.9) \qquad \alpha = A(F,\rho,G,S,M).$$

Note in particular that based on previous assumptions, $A(\cdot)$ is strictly concave in the variables under the firm's control (F, $\rho$, and S) and has the following derivative signs:

$$\partial A/\partial F > 0$$

$$\partial A/\partial \rho \gtrless 0$$

$$\partial A/\partial S > 0 \quad \text{for } S < S_o$$
$$\partial A/\partial S < 0 \quad \text{for } S > S_o$$

$$\partial^2 A/\partial F^2 < 0$$

$$\partial^2 A/\partial \rho^2 < 0$$

$$\partial^2 A/\partial S^2 < 0.$$

Figure 3.5 illustrates the entire R&D production process under the assumption that the firm engages in both a private R&D process and a

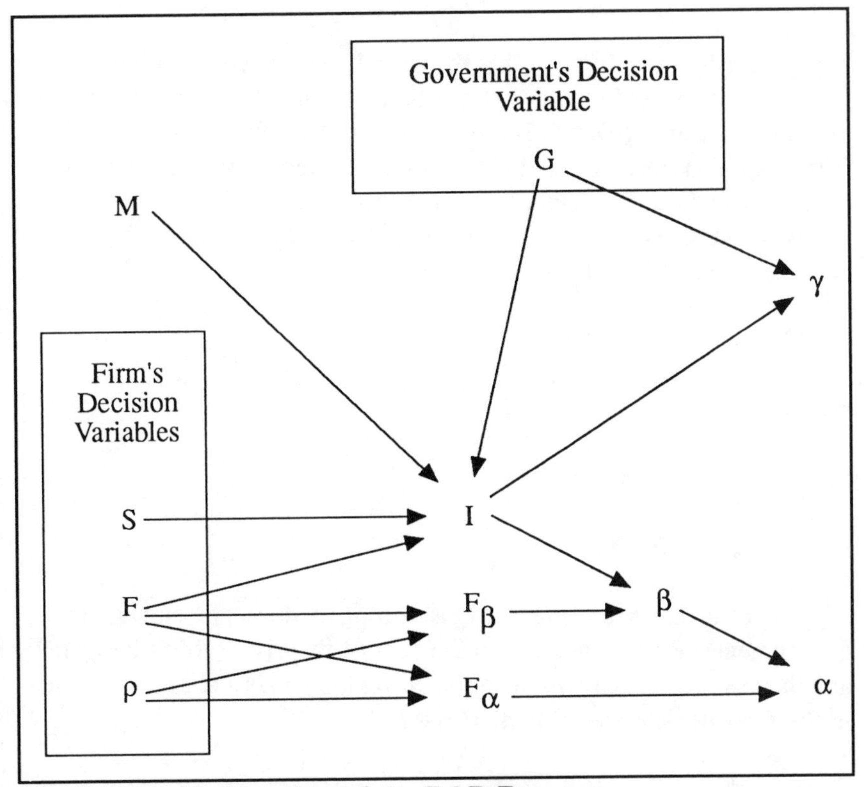


**Figure 3.5. An Overview of the R&D Process**

governmental R&D process.

## Decisionmaking With No Governmental R&D Allocation

In the absence of a governmental R&D allocation to a private-sector firm, the firm and the government will act independently. For the firm, this independence means (1) maximizing profits through the choice of an R&D budget; (2) dividing the R&D budget between basic research, and applied research and development; and (3) determining the level of sharing. For the government, this independence means conducting the governmental R&D process directly and without the benefit of the private-sector firm's infratechnology.

**The Firm's Problem.** The firm produces an array of goods and services that are sold in both intermediate and final markets. We assume that the associated revenue is a positive, strictly-concave function of the firm's applied research and developmental knowledge, $\alpha$, which, as noted in equation (3.9) embodies the effects of S and M. Increases in $\alpha$ are assumed to increase revenue through its impact on the quality and/or the size of the firm's product line. Thus:

$$R = r(\alpha) \qquad (3.10)$$

such that:

$$\partial r/\partial\alpha > 0$$

$$\partial^2 r/\partial\alpha^2 < 0.$$

Total costs for the firm, assuming it does not engage in the governmental R&D process, are the sum of its private R&D budget, F, and the cost associated with manufacturing the goods and services it sells in the intermediate and final markets:

$$C = F + c(\alpha). \qquad (3.11)$$

We assume that manufacturing costs fall at a diminishing rate with increases in $\alpha$:

$$\partial c/\partial\alpha < 0$$

$$\partial^2 c/\partial\alpha^2 > 0.$$

Profits, the difference between equations (3.10) and (3.11), can thus be represented by the following convex function:

$$\Pi = r(\alpha) - F - c(\alpha). \qquad (3.12)$$

Figure 3.6 provides an illustration of the patterns of revenues, costs, and profits as $\alpha$ increases.

Using equation (3.9), profits can be rewritten as a function of F, $\rho$, S, and M as follows:

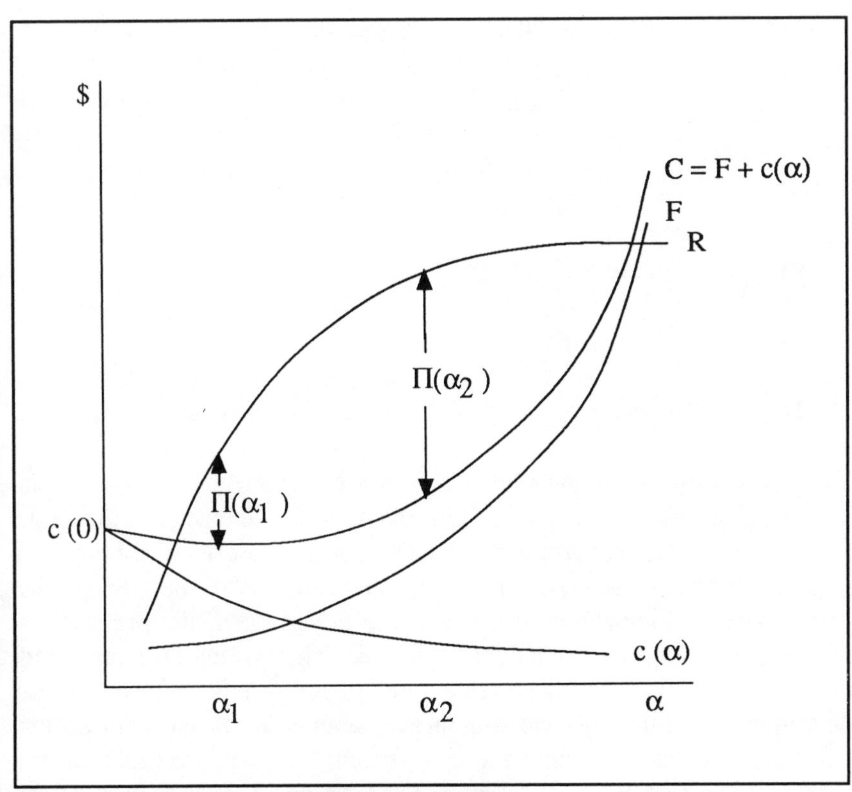


**Figure 3.6. Revenues, Costs, and Profits**

(3.13) $$\Pi = R(F,\rho,S,M) - F - C(F,\rho,S,M).$$

Note in particular that G does not enter equation (3.13) because it is zero.

Because M is exogenous, the firm's problem is to maximize profits over the decision variables F, $\rho$, and S. See Figure 3.7. This problem is strictly concave in the decision variables. Hence, the solution to the firm's problem is unique and characterized by the following three first-order conditions:

(3.14) $$\partial\Pi/\partial F = \partial R/\partial F - 1 - \partial C/\partial F = 0$$

(3.15) $$\partial\Pi/\partial\rho = \partial R/\partial\rho - \partial C/\partial\rho = 0$$

$$(3.16) \qquad \partial\Pi/\partial S = \partial R/\partial S - \partial C/\partial S = 0.$$

These first-order conditions implicitly define the optimal choices for the firm:

$$(3.17) \qquad F' = F(\rho',S',M)$$

$$(3.18) \qquad \rho' = \rho(F',S',M)$$

$$(3.19) \qquad S' = S(F',\rho',M).$$

Let $\Pi'$ represent the expected profit associated with $(F',\rho',S')$.

**The Government's Problem.** Let the government be interested in acquiring goods whose production requires the input of knowledge derived from the governmental R&D process described above. In general, the determination of the demand for $\gamma$ will be tied to the particular decisionmaking structure of the government, the preferences of those individuals participating in that decisionmaking structure, and the reward structure for those same individuals.[8] We do not explore the determination of that demand structure in detail here. However, it seems reasonable to us to assume the government seeks to maximize some objective function $W(\cdot)$ whose arguments include $\gamma$ as well as a set of other factors, $\mathbf{X}$, tied to the characteristics of governmental structure discussed above.

Because the government has chosen not to contract with the private-sector firm for the production of $\gamma$, the maximization of $W(\cdot)$ is constrained by the technology available to the government for engaging in the governmental R&D process. While the government may or may not have access to the same technology as the private-sector firm, it typically finds that the private-sector's technology has an absolute cost advantage over the technology available to the government. Hence, we assume that the technology available to the government is weakly dominated by the technology available to the private-sector firm. As a result, the production function for $\gamma$ available to the government can be defined as the strictly-concave function:

$$(3.20) \qquad \gamma = \gamma_G(G)$$

where $\gamma_G(\cdot)$ is weakly dominated by the production function available to

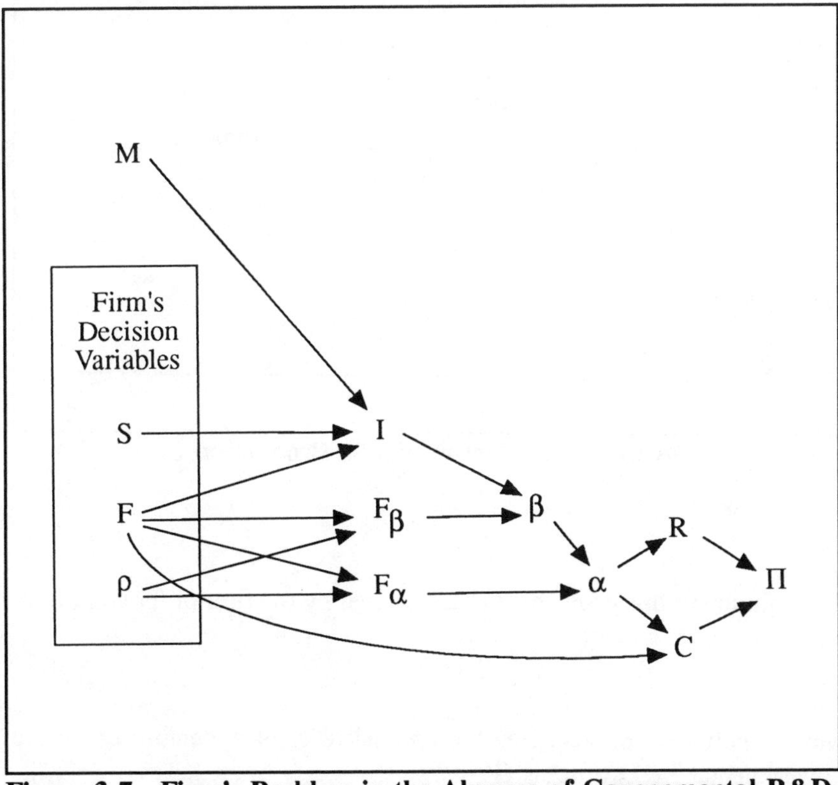


**Figure 3.7. Firm's Problem in the Absence of Governmental R&D**

the private-sector firm:

$$\gamma_G(G) \leq \gamma(G, I).$$

Figure 3.8 illustrates the difference between the production capabilities of the government versus that of the private-sector firm for the case in which the private-sector firm has an absolute cost advantage.[9]

The government's problem, then, is to maximize its objective function $W(\cdot)$ under the production constraint represented by equation (3.20):

$$(3.21) \qquad \max_{G} W(\gamma_G(G), X).$$

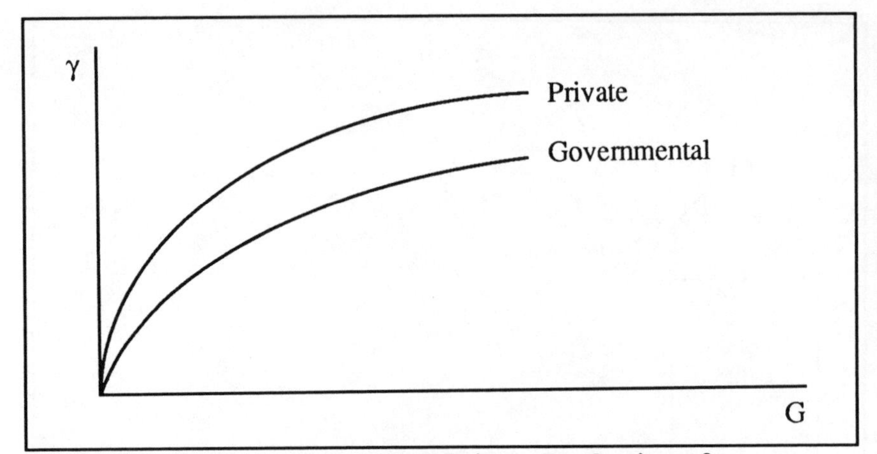


**Figure 3.8. Governmental Versus Private Production of $\gamma$**

The solution to this problem is characterized by the first-order condition:

$$\partial W/\partial \gamma \cdot \partial \gamma_G/\partial G = 0 \tag{3.22}$$

which implicitly characterizes the optimal level of expenditures on the governmental R&D process G′:

$$G' = G(\mathbf{X}). \tag{3.23}$$

Figure 3.9 represents the government's overall problem in the absence of an R&D allocation to the firm.

**Equilibrium.** We assume the government and the firm reach an equilibrium through a Nash equilibrating process. Without the linkage of a governmental R&D allocation to the firm, the Nash equilibrium of the model is trivially defined by the four-equation system consisting of previous equations (3.17)-(3.19) and (3.23), restated and renumbered below as:

$$F' = F(\rho', S', M) \tag{3.24}$$

$$\rho' = \rho(F', S', M) \tag{3.25}$$

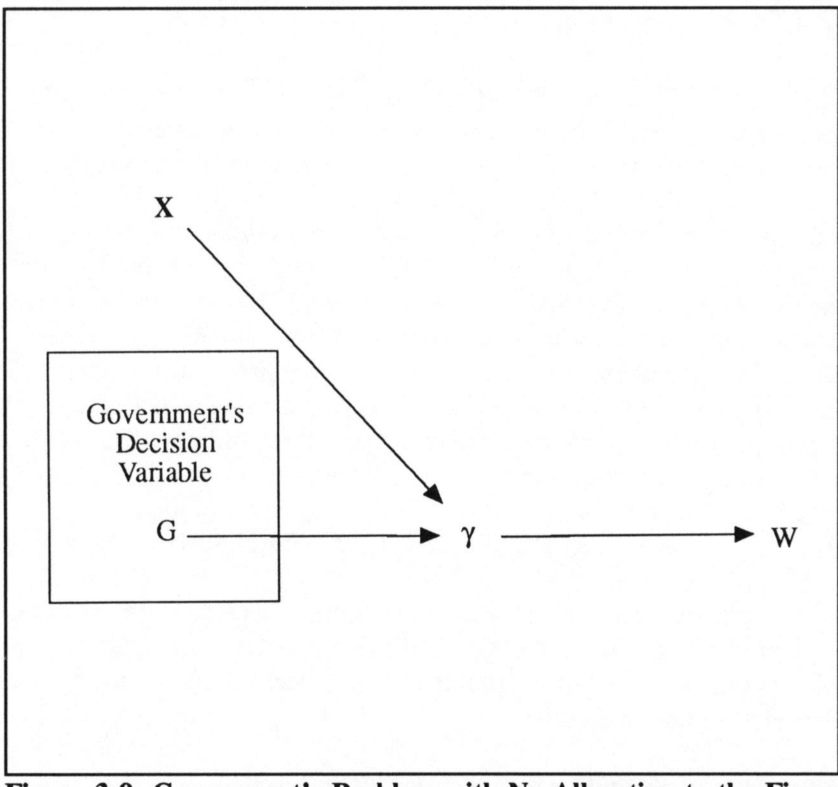


**Figure 3.9. Government's Problem with No Allocation to the Firm**

(3.26) $S' = S(F',\rho',M)$

(3.27) $G' = G(\mathbf{X})$.

## Decisionmaking With a Governmental R&D Allocation

In the presence of a governmental R&D allocation, governmental and firm decisionmaking become connected. For the firm, the governmental allocation results in an increase in the relative level of infratechnology, and therefore a chance to increase profits. For the government, the presence of an R&D allocation to a private-sector firm reflects the judgement by the government that its objective function $W(\cdot)$

can be increased by having the firm do the production.

**The Firm's Problem.** Profits for the firm with a governmental R&D contract will differ from those in equation (3.13) for several reasons. The direct effect on profits of the R&D allocation is an increase in the revenues and the costs of the firm by the amount G.[10] There is, in addition, an indirect effect on profits with a more substantive effect. The relative level of infratechnology available to the firm is assumed to be a positive function of the total amount of funds devoted to all R&D processes conducted within the firm (see equation (3.5)). Hence, because of the governmental R&D allocation, the firm is able to increase the relative level of infratechnology. This results in an enhanced ability to produce $\beta$ and $\alpha$, and therefore to enhance the profit potential for the firm. Thus, the profit function must be redefined as:

$$\Pi^* = R^*(F,\rho,G,S,M) - F - C^*(F,\rho,G,S,M). \tag{3.28}$$

Like the firm's problem in the absence of an R&D contract, this problem is strictly concave in the firm's decision variables. Hence, the solution to this revised firm's problem is unique and characterized by the three first-order conditions:

$$\partial\Pi^*/\partial F = \partial R^*/\partial F - 1 - \partial C^*/\partial F = 0 \tag{3.29}$$

$$\partial\Pi^*/\partial\rho = \partial R^*/\partial\rho - \partial C^*/\partial\rho = 0 \tag{3.30}$$

$$\partial\Pi^*/\partial S = \partial R^*/\partial S - \partial C^*/\partial S = 0, \tag{3.31}$$

which implicitly define the optimal choices for the firm:

$$F^* = F(\rho^*,S^*,G,M) \tag{3.32}$$

$$\rho^* = \rho(F^*,S^*,G,M) \tag{3.33}$$

$$S^* = S(F^*,\rho^*,G,M). \tag{3.34}$$

Let $\Pi^*$ represent the profit associated with $(F^*,\rho^*,S^*)$. A schema of the firm's problem with an R&D contract is shown in Figure 3.10.

It is important to note that the firm will never refuse an R&D contract as characterized in this model. Because the only expected effect

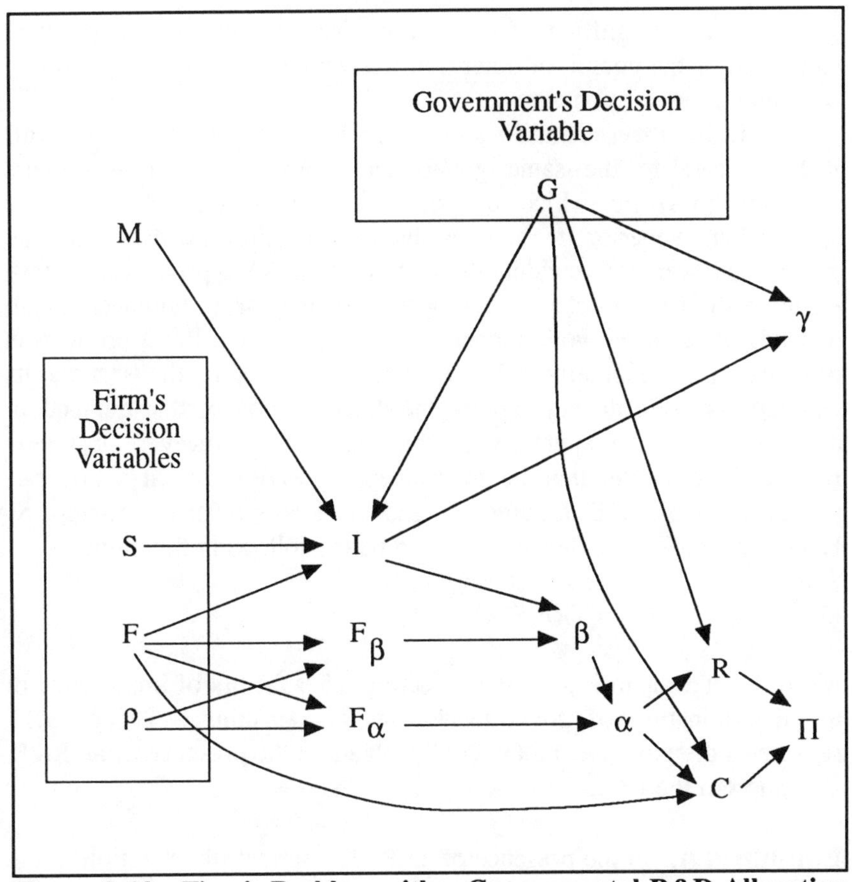


**Figure 3.10. Firm's Problem with a Governmental R&D Allocation**

of the contract is to increase the relative level of infratechnology of the firm at no net cost, the profits associated with engaging in an R&D contract will be greater than the profits associated with refusing an R&D contract, that is $\Pi^* > \Pi'$.

**The Government's Problem.** If the government contracts with a private firm to produce $\gamma$, it does so because it can increase the value of its objective function $W(\cdot)$ over the value associated with producing $\gamma$ itself; that is, because $W^* > W'$. In general, the reasons for giving a governmental R&D allocation to a private-sector firm include the exploitation of some cost advantage that the firm has over the government, the shifting of production risks, the shifting of political risks, and/or the pursuit of other non-R&D objectives such as providing economic rents for constituencies.

If the government does create an R&D allocation for the firm, it has access to the same governmental R&D production process (equation (3.3)) to which the firm has access and which is, by assumption, superior to the production function available to the government were it to conduct the governmental R&D process itself (see equation (3.20)). In characterizing the demand for $\gamma$, however, much depends on what the government knows of the firm's R&D production processes and the amount of R&D activity conducted by the firm and its competitors. We do not explore the determination of that demand in detail (but see the appendix to this chapter). However, it seems reasonable to assume that the government is aware of the firm's overall private level of R&D funding, F, and the firm's level of sharing, S. Hence, we define the demand for $\gamma$ to be the following function:

$$G^* = G(F,S,\mathbf{X}) \tag{3.35}$$

where **X** once again is a vector reflecting other factors of importance in the determination of governmental decision making. Figure 3.11 represents the government's overall problem in the presence of an R&D allocation to the firm.

**Equilibrium.** In the presence of an R&D contract, the solution to the firm's problem will be conditional on the solution to the government's problem and vice versa. The Nash equilibrium of the model is therefore defined by the four-equation system consisting of equations (3.32)-(3.34) and (3.35) as restated and renumbered below:[11]

$$F^* = F(\rho^*,S^*,G^*,M) \tag{3.36}$$

$$\rho^* = \rho(F^*,S^*,G^*,M) \tag{3.37}$$

$$S^* = S(F^*,\rho^*,G^*,M) \tag{3.38}$$

$$G^* = G(F^*,S^*,\mathbf{X}). \tag{3.39}$$

**Explaining Observed Complementarity.** As noted in the previous

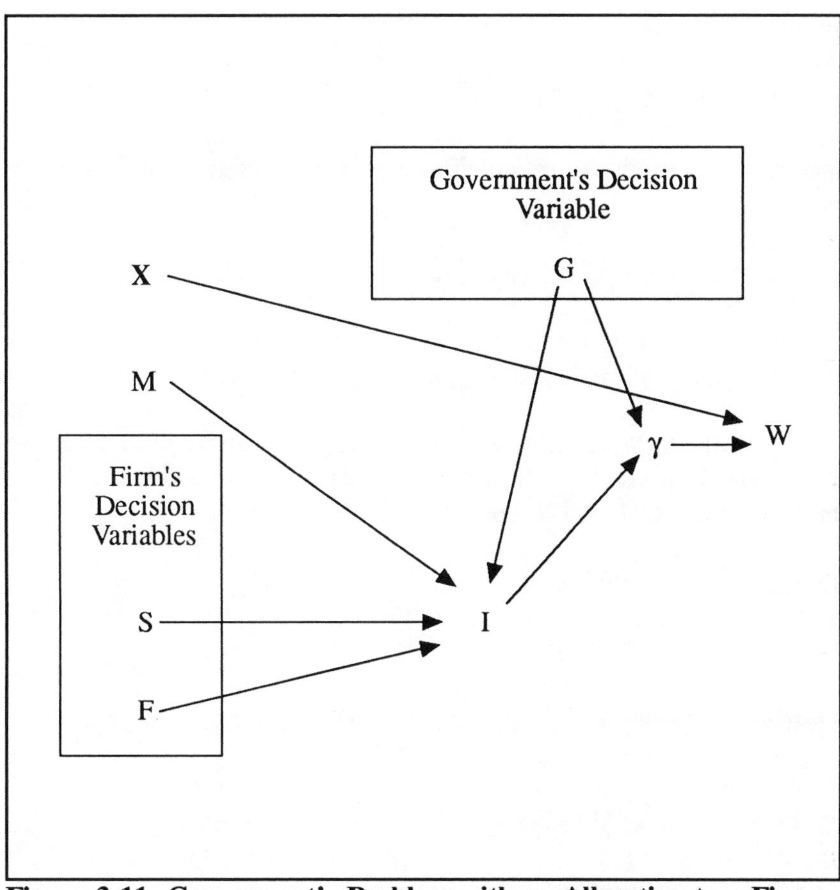


**Figure 3.11. Government's Problem with an Allocation to a Firm**

chapter and highlighted in the introductory paragraphs to this chapter, there is strong empirical evidence of a complementary relationship between private and governmental R&D. The model developed above allows us to trace that complementarity to the presence of technical complementarity in the production of basic knowledge, applied knowledge, and infratechnology.

Two inputs to a production process are technically complementary if a rise in one increases the marginal product of the other. Thus, one input enhances the productive capabilities of another input. Within the context of the model, technical complementarity will exist in the production of $\beta$ and $\alpha$ if the following conditions hold:

(3.40) $\partial^2\beta/\partial F_\beta \partial I > 0$

(3.41) $\partial^2\alpha/\partial F_\alpha \partial\beta > 0,$

and it will exist in the production of infratechnology if the following conditions hold:

(3.42) $\partial^2 I/\partial(F+G)\partial S > 0$

(3.43) $\partial^2 I/\partial(F+G)\partial M_{\text{no sharing arrangement}} < 0$
$\partial^2 I/\partial(F+G)\partial M_{\text{sharing arrangement}} \gtrless 0.$

If the technical complementarity described above does not exist, the model implies the following first-derivative signs for the firm's optimal choice of F and S (equations (3.36) and (3.38)):

(3.44) $\partial F/\partial S \leq 0$

(3.45) $\partial F/\partial G \leq 0$

(3.46) $\partial F/\partial M_{\text{no sharing arrangement}} > 0$
$\partial F/\partial M_{\text{sharing arrangement}} \gtrless 0$

(3.47) $\partial S/\partial F \leq 0$

(3.48) $\partial S/\partial G \leq 0$

(3.49) $\partial S/\partial M_{\text{no sharing arrangement}} > 0$
$\partial S/\partial M_{\text{sharing arrangement}} \gtrless 0.$

The negative inequality noted in condition (3.44) arises because a rise in S (by increasing $\beta$) reduces the marginal profit associated with F. Hence, the firm cuts back on F. In essence, S is an alternative input to F in the production process. Similar arguments explain the inequalities noted in (3.45), (3.47), and (3.48). The effect of M noted in (3.46) and (3.49) depends on whether the firm engages in a cooperative sharing arrangement. If the firm does not engage in a cooperative sharing arrangement, then a rise in M creates a negative externality for the firm by reducing the firm's relative level of infratechnology. Hence, the firm chooses to increase F (or S) to compensate. If there is such a cooperative sharing arrangement, a rise in M may also provide a mitigating positive externality. Hence, the combined effect is ambiguous.

If technical complementarity is present in the production of basic knowledge and in the production of infratechnology (that is, conditions (3.40)-(3.43), hold), then a counteracting increase in the marginal productivity of F (or S) will exist. If the technical complementarity is sufficiently strong, the conditions noted in the inequalities (3.44)-(3.49) will reverse sign. Thus, within the context of the theoretical framework developed above, the assumption of technical complementarity would seem necessary to explain the complementarity often observed between private and public R&D.

However, though technical complementarity may exist, it may be too small to reverse the signs noted in the inequalities (3.44)-(3.49). Thus, the signs noted in those same inequalities are consistent both with the hypothesis that there is technical complementarity and with the hypothesis that there is not any technical complementarity. Clearly, then, direct evidence of technical complementarity requires finding signs opposite to those noted in equations (3.44)-(3.49).

There is a practical importance to understanding these relationships. Public policy toward innovation is now tending to emphasize cooperative research at the basic end of the R&D spectrum and the transfer of basic technologies from Federal laboratories to private sector firms. Clearly, the value of such policies is enhanced if there is technical complementarity.

## EMPIRICAL EVIDENCE

### Descriptive Analysis

Our empirical analysis is based on a 1987 data set of 137 R&D laboratories and is derived from an earlier survey-based effort designed to quantify R&D activity in U.S. industrial laboratories.[12] Hereafter, this data set is referred to as the survey data. For each R&D laboratory, seven variables were constructed. See Table 3.1 for descriptive statistics.

The three primary variables of concern for which data were available are private R&D, sharing effort of the laboratory, and governmental R&D. No data were available on the percentage distribution of private R&D between basic research, and applied research and development.

F represents the laboratory's total private-R&D budget. For the sample of 137 laboratories, the mean value of self-financed R&D is over $23 million. However, as Table 3.1 reveals, there is considerable variation in the amount of self-financed R&D.

S represents the sharing efforts of each R&D laboratory and is roughly approximated from our data set by the percentage of R&D person-hours devoted to activities having public-good characteristics. To construct S, we took data from an original survey question in which laboratory directors were asked to quantify the R&D output of the laboratory by distributing the total number of person-hours among each of the following eight activity categories:

**Table 3.1. Descriptive Statistics (n=137)**

| *Variable* | *Description* | *Mean* | *Standard Deviation* | *Range* |
|---|---|---|---|---|
| F($M) | Private R&D | 23.44 | 72.74 | 1.4-600 |
| S(%) | Relative sharing effort | 8.96 | 12.11 | 0-35 |
| G($M) | Governmental R&D | 3.44 | 19.92 | 0-174 |
| M | R&D-intensive industry | 0.50 | 0.50 | 0/1 |
| K | Basic research | 0.12 | 0.32 | 0/1 |
| BL | Biological / chemical laboratory | 0.56 | 0.49 | 0/1 |
| CR | Cooperative research | 0.49 | 0.50 | 0/1 |

- published articles and books
- patents and licenses
- algorithms and software
- internal technical and scientific reports
- prototype devices and materials
- papers for presentation at external conferences
- demonstration of technological devices
- other products.

S was then defined to be the percentage of each laboratory's time devoted to published articles and books and to papers for presentation at external conferences. These two categories reflect, in our opinion, the output activities that correspond best to the concept of shared technical knowledge.[13] S has a sample mean value of 8.96 percent, though values ranged widely from 0 percent (eight observations) to 35 percent.

G represents the government's total spending to acquire governmental technological knowledge. G is calculated from the survey data to include direct governmental R&D appropriations, contracts or grants, as well as the value of the scientific and technical equipment and facilities financed directly by governmental resources. The sample mean value of G is $3.44 million, though again there was considerable variation with values ranging from $0 (ten observations) to $174 million.

Correlations between F, S, and G are reported in Table 3.2. They indicate (as expected, given previous studies) a high correlation (0.926) between F and G for the sample. In all cases, these correlation coefficients are significant at standard levels of significance. The presence of positive correlation coefficients provides at least preliminary evidence of the existence of technical complementarity at the production level. Correlations between F and S and between G and S are also positive and significant, thus providing further initial evidence of technical complementarity at the production level.[14]

We also focused our attention on four other variables of interest. M represents the R&D efforts of competitors and was approximated by a binary variable equalling unity if the laboratory's principle research area corresponds to an R&D-intensive industry, and zero otherwise. We defined R&D intensive industries, based on NSF-reported R&D to sales ratios, to be chemicals, petroleum refining, non-electrical machinery, electric and electronic equipment, transportation equipment, and instruments. Half of the laboratories contained in the survey data have areas of principle research in an R&D-intensive industry. The three remaining variables are dummy variables taking the value of 0 or 1 and were constructed to mark the presence of important activities for which more detailed data were not available. CR marks the presence of formal inter-laboratory agreements and was intended to proxy for the presence of cooperative sharing arrangements. K marks the presence of basic R&D activity, and BL represents the research focus of the laboratory by noting whether the firm engages in substantial biological or chemical research.[15]

## The Empirical Model[16]

In order to analyze the data further, we sought to parameterize the theoretical framework summarized by equations (3.36)-(3.39) above. However, we were hampered by the inability to construct a suitable measure of $\rho$, the proportion of private R&D funding going to basic

**Table 3.2. Sample Correlations Between Key Variables**

| | F | G | S |
|---|---|---|---|
| F | 1.0 | | |
| G | 0.926<br>(0.0001) | 1.0 | |
| S | 0.220<br>(0.01) | 0.248<br>(0.003) | 1.0 |

*Note:* Levels of significance in parentheses.

research from the sample data set. As a result, some changes in the theoretical framework were required before an empirical model could be estimated.

In modifying the theoretical framework, we noted that $\rho$ had no effect on the level of infratechnology (which was a function of total private R&D funds, F) and hence had no effect on the governmental R&D process. Moreover, $\rho$ had no direct effect on revenues or costs. As a result, we could collapse the basic research process and the applied research and development process into a single private R&D process without disturbing the remainder of the model. Essentially, we created a reduced-form characterization of the private R&D process.

Assuming that the firm receives a governmental R&D allocation, the resulting reduced-form production function available to the firm (see Figure 3.12) thus becomes the strictly-convex function:

$$\alpha = A(F,G,S,M) \tag{3.50}$$

with the following derivatives:

$$\partial A/\partial F > 0$$

$$\partial A/\partial S > 0 \text{ for } S < S_o$$
$$\partial A/\partial S < 0 \text{ for } S > S_o$$

$$\partial^2 A/\partial F^2 < 0$$

$$\partial^2 A/\partial S^2 < 0.$$

$A(\cdot)$ then enters the profit function in the same manner as the $A(\cdot)$ function so that profits for the firm engaged in both private and governmental R&D can be reexpressed as:

$$\Pi^* = R^*(F,G,S,M) - F - C^*(F,G,S,M). \tag{3.51}$$

The optimal levels of F and S for the firm will then be:

$$F^* = F(S^*,G,M) \tag{3.52}$$

$$S^* = S(F^*,G,M). \tag{3.53}$$

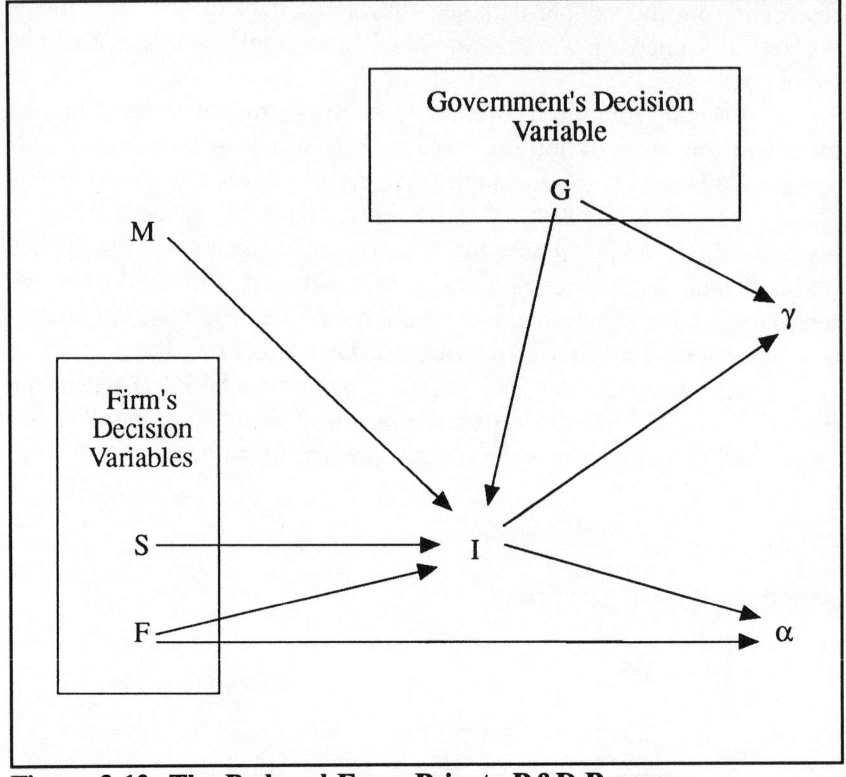


**Figure 3.12. The Reduced-Form Private R&D Process**

The government's demand function stated in equation (3.39) remained the same.

To perform the regression analysis, we parameterized this modified equilibrium system of equations (3.52), (3.53), and (3.39) by the following linear, econometric specification:

$$F = \phi_0 + \phi_1 S + \phi_2 G + \phi_3 M + \epsilon_F \qquad (3.54)$$

$$S = \xi_0 + \xi_1 F + \xi_2 G + \xi_3 M + \xi_4 BL + \epsilon_S \qquad (3.55)$$

$$G = \eta_0 + \eta_1 F + \eta_2 S + \eta_3 K + \eta_4 CR + \epsilon_G. \qquad (3.56)$$

The error terms are assumed to be normally and independently distributed.

Based on the discussion of technical complementarity, the predicted signs for $\phi_1$, $\phi_2$, $\xi_1$, and $\xi_2$ are theoretically ambiguous. However, it was our expectation that some, if not all, of them would take on positive values, thus providing direct evidence of the presence of technical complementarity in both the production of infratechnology and the production of private technological knowledge. $\phi_3$ and $\xi_3$ measure the effect of competitors' R&D effort on the firm's private R&D and sharing efforts. Because some laboratories engaged in cooperative sharing arrangements, the signs of these parameters are theoretically ambiguous. However, our expectation was that the value of a cooperative sharing agreement would be to mitigate only partially the negative effects of a firm's competitors who are engaged in R&D. Hence, we expected $\phi_3$ and $\xi_3$ to be negative. The variable BL was included in equation (3.55) to allow for what we felt were substantial differences in the research culture of biological/chemical R&D versus other industries. Because of our belief that biological/chemical R&D researchers share more information than non-biological/chemical R&D researchers (mostly engineering R&D researchers in our sample), our expectation was that $\xi_4$ would be positive.

We did not predict the signs of $\eta_1$ and $\eta_2$ from the theoretical framework. However, because we suspected that governmental allocations may favor organizations whose research enriches more directly the Nation's science base, we added the variable K to the specification of equation (3.56) with the expectation that its parameter, $\eta_3$, would be positive. Finally, previous work has shown that cooperative sharing arrangements tend to be focused on technologies that are transferred easily between laboratories or firms and, therefore, such organizations might be more likely to receive marginal governmental monies if such funding is done so as to increase overall innovative efficiency.[17] We therefore included CR in equation (3.56) with the expectation that $\eta_4$ would be positively signed.

## The Empirical Results

Equations (3.54)-(3.56) were estimated using two-stage least squares (2SLS). The 2SLS results are reported in Table 3.3.[18] While caution must be exercised in interpreting our findings owing to the exploratory manner in which we empirically approximated several theoretical concepts, the results are generally consistent with the hypothesis that technical complementarity is present in the R&D process. These results also confirm the overall complementarity between private and public R&D observed by previous studies. Of the four coefficients associated with the key variables F, S, and G in equations (3.54) and (3.55), three are positive (two significantly). Thus, there is direct evidence of technical complementarity in the production of private R&D and in the production infratechnology. Sharing increases the productivity of private R&D in producing infratechnology, and infratechnology increases the productivity of private R&D in producing private technological knowledge.

The other parameters, whose signs were derived from the theoretical framework, are also in keeping with our expectations. The estimated values of the coefficients associated with M in equations (3.54)

**Table 3.3. 2SLS Estimates of Equations (3.54)-(3.56)**

Equation (3.54)

$$F = \underset{(2.95)}{19.91} + \underset{(0.05)}{0.05}\,S + \underset{(2.71)}{2.74}\,G - \underset{(-1.79)}{12.53}\,M$$

$R^2 = 0.293$

Equation (3.55)

$$S = \underset{(2.90)}{9.42} - \underset{(-1.54)}{0.33}\,F + \underset{(2.12)}{1.04}\,G - \underset{(-0.80)}{3.93}\,M + \underset{(2.10)}{9.90}\,BL$$

$R^2 = 0.095$

Equation (3.56)

$$G = \underset{(-1.38)}{-2.75} + \underset{(2.45)}{0.22}\,F - \underset{(-0.53)}{0.27}\,S + \underset{(0.63)}{2.28}\,K + \underset{(1.95)}{6.63}\,CR$$

$R^2 = 0.282$

*Note:* Asymptotic t-statistics are in parentheses. $R^2$s are calculated following McElroy (1977).

and (3.55) are negative with one of them marginally significant.

Finally, our suspicions that biological/chemical R&D laboratories share more and that governmental R&D levels are positively affected by the presence of basic research and cooperative sharing arrangements were confirmed (with varying degrees of significance) by the positive signs for $\xi_4$, $\eta_3$, and $\eta_4$.

Overall, these results are additionally noteworthy because they shed light on the mechanism by which governmental R&D stimulates the private R&D funding decision. As mentioned previously, numerous researchers have reported a positive relationship between governmental R&D and private R&D using single-equation models estimated with cross-firm or cross-industry data. The positive and significant coefficient on G in equation (3.54) confirms this finding with our sample of laboratories -- governmental R&D directly stimulates private R&D. Moreover, governmental R&D also stimulates sharing which, in turn, stimulates private R&D. After accounting for the various feedback effects that link the determination of private funding, F, and the degree of sharing, S, we find that a $10 million exogenous increase in governmental R&D would result in a $27.5 million increase in private R&D.[19]

Our results also suggest that governmental R&D affects the composition of an R&D laboratory's output. Greater governmental allocations are associated with a greater sharing of technical knowledge. We would, for example, expect a $10 million exogenous increase in governmental R&D to stimulate an almost 15 percent increase in the amount of time spent on activities associated with publishing articles and books and with presenting papers at external conferences. For an average employee working 2000 hours a year, this would amount to an increase from 179 hours a year to 205 hours a year engaged in such activity. This finding is not inconsistent with the view that government allocates its R&D resources in such a way as to increase social welfare by increasing knowledge *per se*.

## CONCLUSIONS

This chapter has investigated the mechanism by which public and private R&D interact. Using a theoretical framework of R&D activity, infratechnology was shown to be the critical link between the two sources of funding. Observed complementarity between public and private R&D

funding was then linked to the existence of technical complementarity in the production of infratechnology and private technological knowledge.

Using data from a unique sample of industrial R&D laboratories, we found strong evidence using both descriptive and regression analyses to support the claim that private R&D and governmental R&D are complementary. In addition, we found evidence that governmental R&D stimulates the sharing of knowledge outside the laboratory. Finally, we found direct evidence of technical complementarity. Thus, sharing increases the productivity of funds invested in producing infratechnology, and infratechnology increases the productivity of funds invested in the production of private technological knowledge.

There were, to be sure, a number of possible shortcomings associated with the results reported here. Most notable was the difficulty in finding data that reflect the theoretical concepts underlying our theoretical model. Because of the difficulty in finding data on the division of private R&D funds between basic research, and applied research and development, we were forced to estimate a reduced-form version of our original theoretical framework. In addition, we based our empirical testing on a measure of sharing that was constructed from data on the time allocation of R&D scientists within their laboratory. Although our definition of sharing does conform to an intuitive notion of information dissemination, we would have preferred to confirm our results with other measures. Unfortunately, there are no other such data to our knowledge. These data restrictions further limited our ability to incorporate more sophisticated treatments of spillovers and dynamic effects.

Nonetheless, we believe that these results are heartening for three reasons. First, they provide confirmation, using a unique data set and within a new theoretical context, of the existence of complementarity between public and private R&D. Second, they provide a general framework for other researchers to investigate the *mechanisms* by which governmental R&D affects the private sector, rather than simply to investigate the *effects* of governmental R&D. Third, they provide evidence that governmental R&D has general value not only as a stimulator of private R&D, but also as a catalyst for sharing technical knowledge, and that critical to all this is the ability of governmental R&D to increase the infratechnology base of the private-sector firm.

# NOTES

1. Two factors of production are '(technically) complementary' if a rise in one input increases the marginal product of the other input. The opposite case (in which a rise in one input decreases the marginal product of the other input) has several labels. Hicks (1948) refers to it as a 'regressive' situation, while Ferguson (1979) labels such inputs 'competitive' or 'alternative.'

2. This model is an extension of a model developed under the financial support of the National Science Foundation. See Link, Bozeman, and Leyden (1990).

3. See Link and Tassey (1987) for a description of the general place of infratechnology in facilitating the R&D process. Our notion of infratechnology is somewhat broader than that characterization.

4. See Link and Tassey (1987) and Link and Bauer (1989) on the issue of knowledge sharing.

5. Note that the values of $\beta$ and I are relative to the general levels of basic knowledge and infratechnology among the firm's competitors. Thus, for example, an increase in the (absolute) level of basic knowledge by competitors, *ceteris paribus*, would result in a smaller $\beta$ for the firm.

6. Applied knowledge produced in one time period may affect the level of basic knowledge produced in some future time period. We have chosen not to incorporate this possibility into the model.

7. See Cohen and Levinthal (1990).

8. There are a number of approaches to modeling the determination of G. One possibility, explored in the appendix to this chapter, is to view government as a risk-averse, budget-maximizing bureaucracy. Under such a structure, government provides private-sector R&D allocations both to exploit lower production costs and to insure itself against possible production and political risks.

9. The appendix to this chapter provides a more detailed explanation of why the private sector is more productive. See the discussion surrounding equation (3A.18) and the conclusion to the appendix.

10. We assume that the firm earns a normal rate of return on the governmental R&D process so that economic profits are zero. Thus, the revenues, G, from the government just cover the costs of production associated with the

governmental R&D production process.

11. Particulars of the contracting process may provide additional constraints on the character of the equilibrium actually attained. Note, for example, the literature on auctions. We do not investigate these effects.

12. This data collection effort is discussed in detail in Bozeman and Crow (1988). The original R&D laboratory data base included 574 industrial laboratories, 405 of which were in the manufacturing sector. See Link, Bozeman, and Leyden (1990) for a discussion of how the final sample was selected.

13. In an effort to place this measure of sharing within the existing R&D literature, we correlated it with some traditional innovation-related variables. Two external data sources were used.

First, 1985 firm-level data from the Census/NSF Longitudinal R&D Data Base were accessed, and data on the following variables were collected: percent of each firm's total R&D budget allocated to process applied research and development; percent of each firm's self-financed R&D allocated to projects whose total life in the R&D cycle is more than five years; and percent of each firm's total R&D funded by the Department of Defense. Our original sample of 137 laboratories was selected based on our ability to match each laboratory's parent firm with data in the Census/NSF Data Base. Given these data, shared technical knowledge, as we measured S, was found to be unrelated to the product/process mix of a firm's R&D. As expected, sharing is less evident in the more basic research firms (simple correlation coefficient -0.118, significant at the 0.20 level). Sharing is also less evident in firms with a larger percentage of their R&D allocated to long-term projects (simple correlations coefficient -0.136, significant at the 0.15 level). Surprisingly, the correlation coefficient between sharing and the percentage of R&D that is defense-related is positive. Perhaps those firms receiving R&D funds from the Department of Defense are obtaining them on a product development contract basis. Thus, the source of funding would not necessarily imply that the research was proprietary. This explanation, however, is conjecture on our part.

We also examined the Levin et al. (1985, 1987) data on appropriability. We related our measure of sharing, averaged across laboratories at the four-digit SIC level, with a Levin et al. (1985) industry-level appropriability index. Across industries, there was no significant relationship between the intensity to which resources are devoted to the sharing of technical knowledge and the use of strategies to appropriate technical knowledge.

Any effort to introduce into the literature a new innovation-related construct, such as our sharing variables, will face criticism and scrutiny. We realize, too, that a case could be made that all of the output activities categorized above becomes public over time. For example, it is possible, over time, to decode software or to reverse engineer prototype devices. Nonetheless, the empirical analysis presented below is based on our original definition of sharing as discussed above and is not the result of an *ad hoc* search to find those categories that verified our model the best.

14. We also examined descriptively the sharing variable, S, by industry, field of research, and level of governmental R&D. In brief, we found that S was larger in R&D-intensive industries. Interestingly, R&D-intensive industries also receive the larger share of governmental R&D.

15. We defined a laboratory as biological/chemical if it had more than 25 percent of its scientific and technical personnel in the relevant research area as reported on the survey. Non-biological/chemical laboratories are generally engaged in research in the engineering fields.

16. The empirical model is an extension of earlier, preliminary work. See Link, Bozeman, and Leyden (1990) and especially Leyden and Link (1991).

17. See Link and Bauer (1989).

18. To determine the statistical merits of our 2SLS results compared to going to a third stage of estimation, we follow Belsley (1988). He suggests (p. 28) that "it would seem safe to say that 3SLS would possess good small-sample relative efficiency [relative to 2SLS] for values of $\lambda_{min}$ and det(**R**) in the neighborhood of 0.1 and for values of $\kappa$(**R**) above 20-30." Here, the determinant of the cross-equation correlation matrix, det(**R**), is 0.60 and the minimum eigenvalue of **R**, $\lambda_{min}$, is 0.38. $\kappa$(**R**) is the ratio of the maximum eigenvalue to the minimum eigenvalue of **R**, and it is 4.42. Thus, following Belsley's guidelines, little efficiency would be gained from going to a 3SLS estimation.

19. To generate the net effect of an exogenous change in G on F and S, Cramer's Rule and the coefficients in equation (3.54) and (3.55) were used.

# 3A
# Government as a Risk-Averse Bureaucrat

Studies of governmental R&D allocations to private-sector firms have generally focused on the effects of such allocations on private-sector firm behavior (see Chapter 2). The motivations for government's involvement, however, has received little attention. Where the issue has been extensively examined,[1] it has typically been from a normative perspective and has not been intended to provide a positive analysis of governmental motivations and the effect of those motivations on the form of R&D contracts.

This appendix provides a positive analysis of the behavior of a risk-averse, budget-maximizing bureaucrat charged with conducting an R&D production process for some sponsor, such as a legislature.[2] Two difficulties arise in conducting this analysis. First, the sponsor is unable to monitor the bureaucrat closely and must therefore rely on an indirect measure of performance. Second, the bureaucrat is exposed to uncertain political and production risks that reduce its ability to conduct the R&D process. Hence, the bureaucrat must choose a specific level of effort given indirect monitoring by its sponsor and uncertainty associated with the outcome of the R&D.

If R&D allocations to private-sector firms are not an option, the bureaucrat, of course, must bear all of the risks. However, if such allocations are allowed, the bureaucrat may prefer to provide a private-sector firm with an R&D allocation rather than to conduct the R&D process itself. The form of the R&D contract with the firm will depend on whether the firm is risk neutral or risk averse. If the firm is risk neutral, the contract will be a fixed-fee contract with guaranteed output. If the firm is risk averse, the R&D contract will be of the cost-plus variety with specific terms dependent on the relative degree of risk aversion attributable to the bureaucrat.

The results presented here are important for at least two reasons. First, they suggest that an increased threat from a bureau's sponsor (such as a legislative committee) will affect the production decisions of the bureaucrat. Joskow (1974), for example, argues that public-utility regulatory agencies modify their policies in order to minimize conflict and criticism. To the extent that threats from sponsors are credible, bureaucrats will be more cognizant of political and production risks and will be more likely to have production performed by the private sector. Second, our analysis sheds light on the determinants of fixed-fee versus cost-plus contracting. Sappington and Stiglitz (1987) argue that if the private sector is relatively more risk averse, rents can be captured by the government to the extent that it bears the risks of production. Similarly, Stiglitz (1988) argues that government's superior ability to spread risk provides a rationale for the cost-plus contracts so common in defense procurements. We find that cost-plus contracts will be preferred over fixed-fee contracts whenever any degree of private-sector risk aversion is present, though the relative degree of risk aversion between the bureaucrat and the private-sector firm will affect the degree to which the bureaucrat shares in any added costs of production.

## GOVERNMENTAL R&D CONDUCTED IN-HOUSE

Following Niskanen (1971), consider a budget-maximizing bureaucrat who received a budget, B, from a legislature in return for conducting an R&D production process. See Figure 3A.1 for a decision time line of this process. Assume, moreover, that the legislature cannot observe the effort of the bureaucrat directly and so must, instead, rely on the observation of the bureaucrat's output $\gamma$.[3] As a result, the legislature makes the budget size a function of $\gamma$:

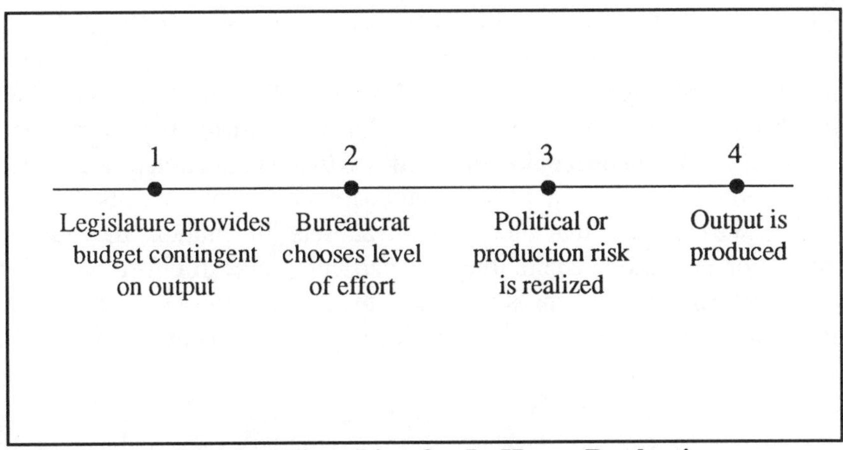


**Figure 3A.1. Decision Time Line for In-House Production**

(3A.1) $$B = B(\gamma).$$

In determining the structure of the budget function B(γ),[4] we assume that the legislature has a preference for a particular quantity of output and is willing to reward further bureaucratic efforts. However, it also wishes to send clear and strong signals that decreased bureaucratic effort will not be rewarded. Given the inability to view effort directly, the legislature structures the budget function such that: [5]

(3A.2) $$dB/d\gamma > 0$$

(3A.3) $$d^2B/d\gamma^2 < 0.$$

Finally, assume that the legislature requires the bureaucrat to balance its budget; that is, total costs C must be equal to the allocated budget:

(3A.4) $$C = B(\gamma).$$

Given the budget, the bureaucrat begins the R&D production process. Bureaucracies are generally exposed to both production and political risks in producing a service. Production risks have their origins in the physical production process itself and are not directly connected to political forces. Their effect is to reduce the productivity of the bureaucrat and therefore to raise the costs of production. Political risks, while also reducing the productivity of the bureaucrat, find their origins in the political arena. For example, Joskow (1974) in his examination of public-utility regulatory agencies, describes the early 1970s as a time characterized by an unanticipated slowdown in the regulation process due to inflation and to the rise of politically active environmentalists.[6]

We assume that the bureaucrat sets in motion the R&D production process by committing to a particular level of effort, and that only after that commitment is made will the production risks or political risks manifest themselves in a particular realization. Thus:

$$\gamma = \gamma(E,\theta) \tag{3A.5}$$

where E represents the level of bureaucratic effort and $\theta$ represents the presence of production or political risk. An index variable, $\theta$, is assumed to be a random draw from a set of possible states-of-nature $\Theta$ and has the known distribution function $f(\theta)$. We let $\theta$ take on non-negative values with greater values of $\theta$ implying increasingly lower values of $\gamma$:

$$\partial\gamma/\partial\theta < 0 \tag{3A.6}$$

$$\partial^2\gamma/\partial\theta^2 \leq 0. \tag{3A.7}$$

For a given level of effort, we define the maximum output level to be that level associated with $\theta$ equal to zero. Hence, $\theta>0$ defines a loss, L, of output:

$$L = L(\theta;E) = \gamma(E,0) - \gamma(E,\theta). \tag{3A.8}$$

Finally, we assume that $\gamma$ is a positive, strictly-concave function of E and that increases in $\theta$ reduce the marginal productivity of E:

$$\partial^2\gamma/\partial\theta\partial E < 0. \tag{3A.9}$$

The cost of producing $\gamma$ is assumed to increase at an increasing rate with the total level of effort expended, that is:

$$C = C(E), \tag{3A.10}$$

such that:

(3A.11) $dC/dE > 0$

(3A.12) $d^2C/dE^2 > 0$.

Hence, the bureaucrat's problem is to maximize the expected ($^e$) budget:

(3A.13) $B^e(E) = \int B(\gamma(E,\theta)) f(\theta) d\theta$

subject to the constraint that the budget balances in expectation:

(3A.14) $\int B(\gamma(E,\theta)) f(\theta) d\theta = C(E)$.

Given the assumptions about the production function for $\gamma$ and the preferences of the legislature as embodied in the budget function, B, this problem has an unique solution characterized by the first-order condition:

(3A.15) $(1+\lambda) dB/d\gamma [\int \partial\gamma/\partial E f(\theta) d\theta] = dC/dE$.

Equation (3A.15) implicitly defines the optimal level of effort $E'$, and thus the optimal expected budget $B'$. However, while the budget balances in anticipation, it may not balance after any given realization of $\theta$. We assume that there is sufficient "slack" in the bureau's operations to accommodate last minute, unanticipated budgetary deficits or surpluses. Anecdotal evidence in the form of end-of-the-fiscal-year buying sprees and belt-tightening measures suggests bureaucrats are quite skilled at such last minute adjustments.

Note, finally, that although the bureaucrat's preferences are linear in the size of the budget, and thus ostensibly risk neutral, the concavity of the budget function offered by the sponsor makes the bureaucrat risk averse. Of course, there are other reasons why bureaucrats might be risk averse. Wilson (1989) suggests that the nature of private versus public enterprises might lead to self selection among individuals such that the more risk averse take jobs in government, for example. Interestingly, while many base their analysis on an assumption of bureaucratic risk aversion (Mueller, 1989), there is little empirical evidence either to support or refute the assumption (Wilson, 1989).[7]

## CONTRACTING WITH A RISK-NEUTRAL FIRM

Because it is risk averse, the bureaucrat has an interest in mitigating the effects of risk by contracting with a private-sector firm for the production of governmental R&D. Define an R&D contract as a promise by the private-sector firm to deliver to the bureaucrat a specified quantity, $\gamma_c$, of the governmental technical knowledge and an insurance payment, I (conditional on the loss L associated with the production of $\gamma_c$), in return for a payment, of G dollars. (Note that I refers to insurance and not to infratechnology as in the main body of Chapter 3.) Assume that the insurance payment, I, is denominated in units of $\gamma$ and equal to some quantity no greater than L:

$$0 \leq I(L) \leq L. \quad \text{(3A.16)}$$

Thus, the contract is composed of a production clause which defines $\gamma_c$ and an insurance clause which defines I. Let $F_c$ be that portion of G which goes for the purchase of $\gamma_c$, let P be that portion of G which goes for the purchase of insurance, and note that:

$$G = F_c + P. \quad \text{(3A.17)}$$

Assume that the firm has the same production technology as the bureaucrat. Assume further that the firm is affected in the same way by production risks but to a lesser degree by political risks. Thus, we can define the firm's production function as $\hat{\gamma}(E,\theta)$ and note in general that:

$$\hat{\gamma}(E,\theta) \geq \gamma(E,\theta). \quad \text{(3A.18)}$$

The level of effort put forth by the firm is a function of the funds devoted to the production process. In order for the effect of differing input costs not to cloud the analysis, we assume that the firm and the bureaucrat have the same cost equation (equation (3A.10)). Hence, the firm's effort will be a positive, concave function of the funds, F, devoted to the production process:

$$E = C^{-1}(F). \quad \text{(3A.19)}$$

The contracted level of output, $\gamma_c$, is defined as the level of output generated by an input of $F_c$ funds under the assumption of no risk, that is, $\theta=0$:

(3A.20) $$\gamma_c = \hat{\gamma}(C^{-1}(F_c),0).$$

Assume that the cost to the firm for administering the insurance portion of the contract is fixed and, for expositional convenience, let it equal zero. Assume also that there is a sufficient degree of competition in the private sector so that the economic profits from the R&D contract are bid to zero. The premium for the insurance portion of the contract will then be actuarially fair and equal to the expected cost of the insurance payment, I. The expected cost to the firm of providing the insurance will depend upon how well the firm meets its contractual obligations despite the loss, L. In reality, a firm would fulfill a contract by making adjustments throughout the production process. Should it become clear that the contracted quantity, $\gamma_c$, will not be produced under current circumstances, the firm, in an iterative fashion, would modify its production process.

To incorporate this dynamic adjustment process into our static model, suppose that the contract allows for the production of $\gamma_c$ to take place in two steps, as suggested in Figure 3A.2. The first step has already been described: decisions are made before $\theta$ is known, some quantity $\gamma_o$ is produced, and a loss, L, equal to $(\gamma_c-\gamma_o)\geq 0$ is realized. If and only if this loss is positive will the second step take place.

Knowing the value of $\theta$ from the first step, say $\theta_o$, we assume the firm must re-perform the production of $\gamma$, augmenting the process with additional private funds, $F_I$, so as to produce an additional quantity, I. Because $\theta$ in this second step is known, the production process is now deterministic. Hence, $F_I$ is implicitly defined by the equation:

(3A.21) $$\gamma_c - L + I = \hat{\gamma}(C^{-1}(F_c+F_I),\theta_o).$$

Thus, given a particular insurance function, I, $F_I$ can be written in general notation as:

(3A.22) $$F_I = F_I(F_c,\theta).$$

Note that $F_I$ is a positive function of $F_c$.

This process is illustrated in Figure 3A.3. Let the bureaucrat contribute $F_c$ dollars in exchange for the contracted quantity $\gamma_c$. In the first stage of production, the firm realizes a loss, $L = \gamma_c-\gamma_1$, as a result

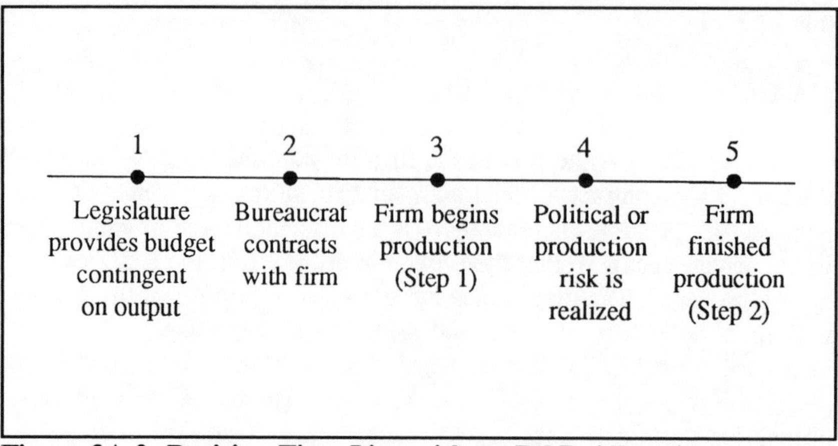


**Figure 3A.2. Decision Time Line with an R&D Allocation**

of drawing a $\theta$ equal to $\theta_1$. Actual total output from stage one is $\gamma_1$ which, as drawn, is less than the contracted quantity $\gamma_c$. A second stage of production is therefore required. Given that the firm now knows $\theta = \theta_1$, the firm determines that $F_I$ dollars are required to increase the total output by I, the contracted insurance payment.

$F_I$ is the actual cost to the firm of providing the insurance payment, I, to the bureaucrat. The actuarially fair premium, P, is therefore the expected value of $F_I$:

$$(3A.23) \qquad P = F_I^e = \int F_I(F_c,\theta)\, f(\theta)\, d\theta.$$

Thus, the premium will be a positive function of the bureaucracy's production costs $F_c$:

$$(3A.24) \qquad P = P(F_c).$$

Finally, note that the expected profits for the firm will be:

$$(3A.25) \qquad \Pi^e = G - F_c - F_I^e$$

which, by equations (3A.17) and (3A.23), equals zero. By construction, the firm is risk neutral.[8]

Given equations (3A.17) and (3A.24), the choice of G uniquely

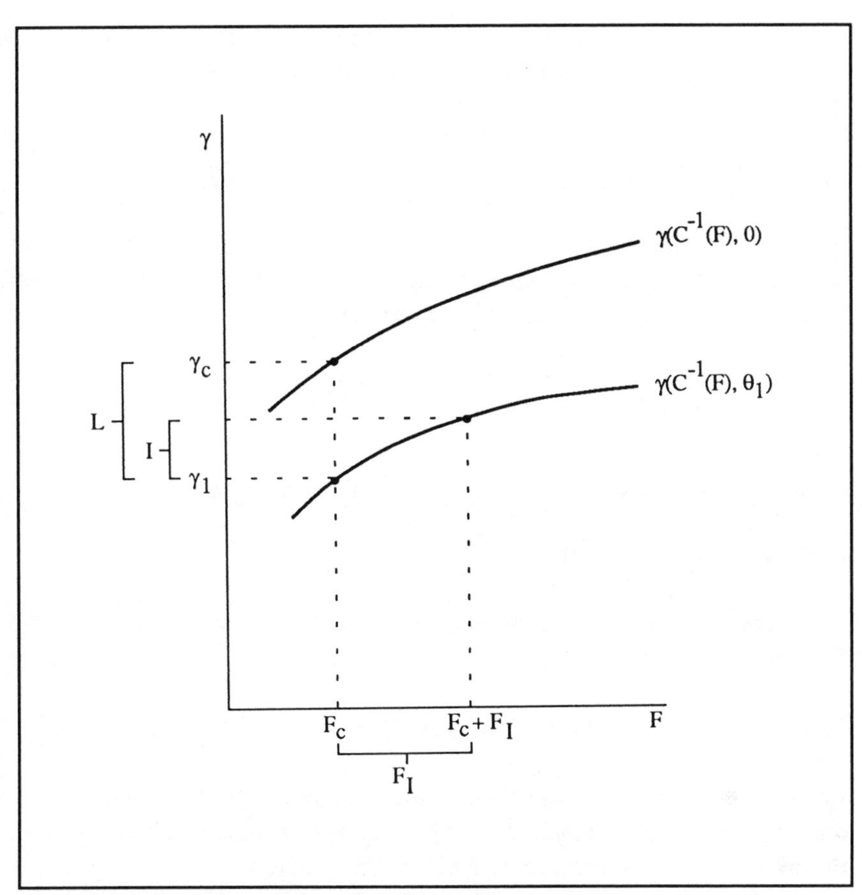


**Figure 3A.3. Determining the Cost of Insurance**

determines the contracted quantity $\gamma_c$. Thus, we rewrite $\gamma_c$ (equation (3A.20)) as:

(3A.26) $$\gamma_c = \gamma_c(G).$$

Note that $\gamma_c$ is a positive, strictly-concave function of G with G the bureaucrat's only decision variable.

The bureaucrat's problem, assuming he contracts with the private-sector firm for the production of $\gamma$, is to maximize the expected size of its budget through the appropriate specification of the insurance function I(L) and the appropriate choice of G. Given the assumption of a risk-neutral firm and a fixed cost of administering the insurance portion of the contract, the bureaucrat will choose an insurance contract with no deductible and no coinsurance, that is:[9]

$$(3A.27) \qquad I = L.$$

Thus, the optimal contract will be a fixed-fee contract with the firm guaranteeing delivery of the quantity $\gamma_c$. The bureaucrat's problem therefore becomes the deterministic problem:

$$(3A.28) \qquad \max_{G} B(\gamma_c(G))$$

subject to the constraint that the budget balances:

$$(3A.29) \qquad B(\gamma_c(G)) = G.$$

The solution to this problem is unique and characterized by the first-order condition:

$$(3A.30) \qquad (1+\lambda)\, dB/d\gamma_c\, d\gamma_c/dG = 1$$

which implicitly defines the optimal level of governmental spending, $G^*$, and thus the optimal budget, $B^*$. Whether, in fact, the bureaucrat prefers to have the R&D conducted by a private-sector firm rather than produce the R&D in-house will depend on the effect of the R&D allocation on budget size. If the R&D allocation increases budget size, that is, if $B^* > B'$, then the bureaucrat will rationally choose to have the firm perform the R&D. To determine whether $B^*$ is greater than $B'$, note first that there is an output level $\gamma_{CE}$ which is lower than $\gamma'$ (the expected output if the R&D is produced in-house) and which, if produced with certainty, would generate a budget equal in size to $B'$. The budget which will obtain under an R&D allocation contract, $B^*$, will be greater than the expected in-house production budget, $B'$, if the R&D allocation with a level of output set equal to $\gamma_{CE}$ results in a budgetary surplus; that is, if:

$$(3A.31) \qquad B(\gamma_{CE}) > G_{CE}$$

where $G_{CE}$ is the cost to the bureaucrat of an R&D allocation with an output of $\gamma_{CE}$. Figure 3A.4 provides an illustration under the assumptions that $\theta$ represents production risks, that $\theta$ can take only a value of 0 or some $\theta_1 > 0$, and that the probability of $\theta = 0$ and $\theta = \theta_1$ are both 1/2. If the bureaucrat conducts the R&D in-house, the solution to the bureaucrat's problem is to expend an effort that generates costs of F′. At F′, there is a 50 percent chance that output will be $\gamma_H$ and a 50 percent chance that output will be $\gamma_L$. Hence, the expected budget will be B′. At B′ the balanced budget requirement is met (B′ equals F′) and the expected level of output is $\gamma'$. Note, finally, that $\gamma_{CE}$ is the level of R&D which, if produced with certainty, would also generate a budget of B′.

If the bureaucrat contracts with a private firm to produce $\gamma_{CE}$ with full insurance, the firm's costs will be $F_c$ with probability 1/2 and $F_1$ with probability 1/2. (For the moment, ignore the curve in Figure 3A.4 labeled $\gamma(C^{-1}(F),\theta_2)$.) Hence, the expected cost to the firm of producing $\gamma_{CE}$ will be $(F_c+F_1)/2$ which is smaller than B′.[10] Thus, there is a budgetary surplus, and because of this surplus it pays to engage the private-sector firm. With the optimal R&D contract, output will be somewhat larger than $\gamma_{CE}$ and therefore B* will be larger than B′. In general, the contracted quantity $\gamma_c$ may be larger or smaller than the expected in-house quantity $\gamma'$. However, if the private firm can produce the quantity $\gamma'$ at a cost less than F′, the level of $\gamma_c$ will also be greater than $\gamma'$.

It is possible, however, that the bureaucrat will prefer the riskier option of in-house production. Burness, Montgomery, and Quirk (1980), for example, examine the decisionmaking of a regulated firm in a Joskow-type model of regulation and note that under rather general circumstances the firm will actually prefer a risky option over a risk-free alternative. Figure 3A.4 illustrates these circumstances within the context of our model. If $\theta$ equals 0 with probability 1/2 and some $\theta_2$ with probability 1/2, and if $\theta_2$ is sufficiently greater than $\theta_1$, we find that while the solution to the bureaucrat's in-house production problem is the same, the expected cost to the firm of producing $\gamma_{CE}$ will be $(F_c+F_2)/2$ which is larger than B′. Hence, an R&D contract for $\gamma_{CE}$ would result in a budgetary deficit, and only a contract for a lower level of output would balance the budget. However, this would reduce the budget size below what the bureaucrat expects to get through in-house production and will therefore not be preferred. The choice of a riskier alternative, therefore, does not imply a preference for risk *per se*. Rather, it indicates that the cost of insuring against the risk exceeds the price the

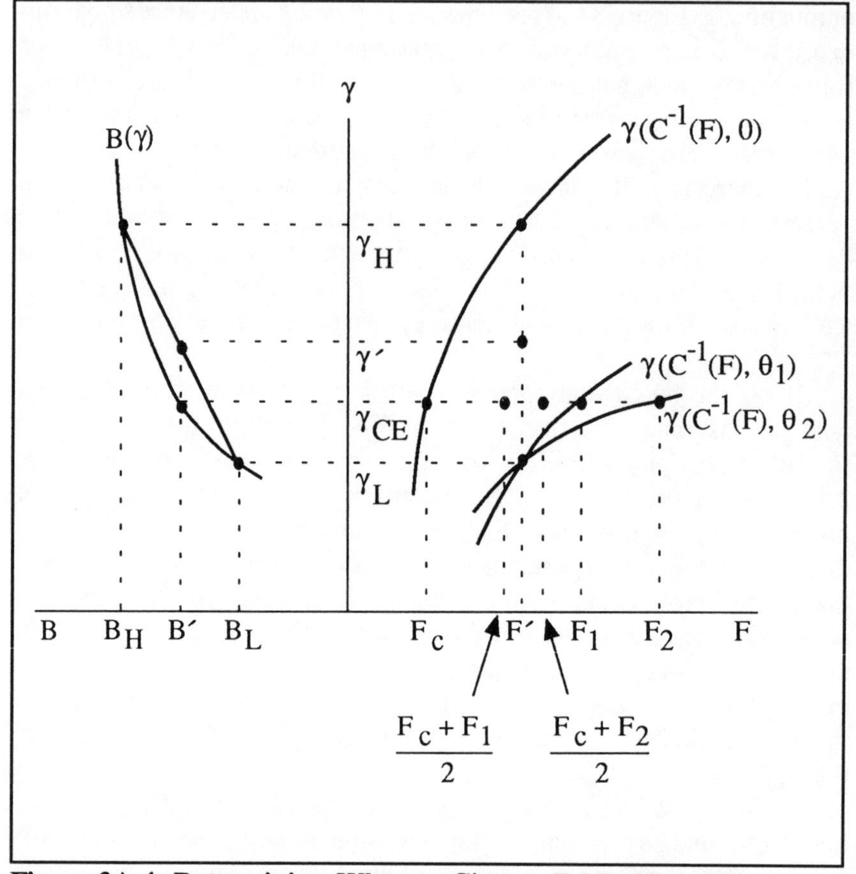


**Figure 3A.4. Determining When to Give an R&D Allocation**

bureaucrat is willing to pay.

This analysis applies all the more if $\theta$ represents political risks. Recall that if the risk is political in nature, the private firm will be affected less by any realization of $\theta$ than the bureaucracy would, that is, $\hat{\gamma} > \gamma$. Hence, the likelihood of giving an R&D allocation to a private-sector firm will be greater because the firm provides an absolute cost advantage in addition to its role as insurer. Note, finally, that the likelihood of a private R&D allocation is directly connected to the budget function provided by the sponsor and is therefore subject to manipulation. Changes in the political climate that lead to a more concave budget function, $B(\gamma)$, due, for example, to greater penalties for production below a target level, will tend to reduce the certainty equivalent budget, $\gamma_{CE}$, and thereby make it more likely that a private R&D contract will be chosen.

## CONTRACTING WITH A RISK-AVERSE FIRM

The analysis so far has assumed that the private-sector firm is risk neutral. However, risk aversion is a possibility for many firms, particularly those that are small or for whom governmental contracts make up a substantial portion of their business.[11] We note that if the private-sector firm is risk averse, the optimal insurance contract will now include a coinsurance clause, that is:[12]

(3A.32) $$I = \alpha L$$

where $\alpha$ represents the degree to which the bureaucrat will share the production risks with the private firm and is determined by the relative degree of bureaucratic risk aversion:

(3A.33) $$\alpha = R_G / (R_G + R_F).$$

$R_G$ represents the index of absolute risk aversion for the bureaucrat and $R_F$ represents the same measure for the firm.[13] Hence, if the bureaucrat insists on a particular quantity of output, it must share in any cost overruns, that is, it must engage in a cost-plus contract.

The bureaucrat's problem is thus to maximize the expected value of:

(3A.34) $$B(\gamma_c(G)-(1-\alpha)L)$$

subject to the constraint that the budget balance in expectation:

(3A.35) $$\int B(\gamma_c(G)-(1-\alpha)L) f(\theta) d\theta = G.$$

The solution to this problem is again unique and characterized by the first-order condition:

(3A.36) $$(1+\lambda) dB/d\gamma [d\gamma_c/dG - (1-\alpha) \int \partial L/\partial G f(\theta) d\theta] = 1.$$

The effect of this will depend on the value of $\alpha$. For values of $\alpha$ close to one, that is, if the firm has relatively little risk aversion, the outcome will be very similar to that described for the case of a risk-neutral-firm. However, as the firm's degree of risk aversion rises, $\alpha$ will fall. Hence, the likelihood will fall that the bureaucrat will find a private R&D contract that outperforms what can be achieved by in-house production.

## CONCLUSIONS

This appendix examined the effect of bureaucratic risk aversion on the decision to engage a private-sector firm to conduct R&D for the government. If the private-sector firm is risk neutral, the bureaucrat will prefer to provide an R&D allocation if it can find a firm willing to produce a quantity of output that generates the same budget as the bureaucrat expects to get through in-house production, and do so at lower cost. The contract itself will be a fixed-fee contract with full insurance. If the firm is risk averse, the ideal contract will have only partial insurance with the decision to conduct R&D in-house hinging on whether the expected budget is greater under private production or in-house production. Essentially, this results in a cost-plus type contract for bureaucracies interested in a particular level of output. Because the form of the budget function affects the likelihood of a bureaucracy preferring an R&D contract, sponsors, through appropriate manipulation of the reward and penalty structure, may be able to induce bureaucracies to send more R&D to private-sector firms.

Why is an R&D allocation to a private-sector firm a potentially beneficial alternative for the bureaucrat? Clearly, risk aversion on the part of the bureaucrat plays a part, for without it the bureaucrat would have no incentive to seek insurance. However, beyond the preferences of the bureaucrat, it is also necessary that the private firm have some form of superior technology. We do not intend to suggest by this that the firm can simply conduct R&D cheaper using fewer inputs, though that may be the case. What we have in mind is a more subtle superiority connected to the ability to revise decisions in the light of new information. As Gomez-Ibanez, Meyer, and Luberoff (1990) note, the public sector is often less flexible than its private sector counterpart because of cumbersome work rules and a tendency to engage in a slow production process that requires one stage to be completed before moving on to the next. This is the essence of our model of bureaucratic versus private-sector production processes. For the bureaucrat, the production process is essentially static with input levels chosen before the effects of risk are realized. For the firm, however, the process is dynamic. While the firm also targets a level of output and commits to a level of inputs, it modifies its decisions as the production process progresses. As a result, while the bureaucrat only comes to know the effects of risk *ex post*, the firm through an iterative process comes to know the effects of risk *ex ante*.[14]

It needs to be emphasized, however, that the opportunity to give the R&D production process to a private-sector firm does not necessarily mean that it is desirable. Beyond the factors discussed in this appendix, there may be other risks such as those associated with public employment considerations that arise only if the bureaucrat chooses to use a private-sector firm. To the extent such risks exist, they will act as a counterbalance to the forces that induce the bureaucrat to privatize.

Finally, our analysis suggests that popular perceptions of bloated and padded governmental contracts may in fact have some validity. A contract involving a risky production process will contain a risk premium paid by the government to the private-sector firm in compensation for accepting the burden of those risks. These added costs are, however, unavoidable unless the bureaucrat can be persuaded to bear more risk. The greater the risk, the greater the premium, and therefore the greater the perceived contract padding. While it is not clear how much of the padding in governmental contracts is due to the necessity of paying an insurance premium, it is clear that there are examples of fixed-fee contracts with complete insurance. Perhaps the clearest examples are contracts that require performance bonds. Performance bonds are devices used to ensure that a production contract is enforced. If the firm reneges on its contract, the government receives a compensating payment. While the amount of the bond is open to negotiation, one could argue that such bonds should be set at a level just sufficient to compensate the bureau for having to make alternative arrangements.[15] Not to do so would only hurt the government because the cost to the firm of the bond is built into the contract price. In essence, then, our analysis suggests that such contracts result in the firm guaranteeing the originally contracted output.

# NOTES

1. See, for example, Dasgupta (1987).

2. This appendix is based directly upon Leyden and Link (1992).

3. Lindsay (1976) argues that a legislature typically monitors a subset of all attributes of the mandated service. As a result, the bureaucrat only focuses its attention on the monitored subset of attributes. The reader may therefore prefer to think of $\gamma$ as a vector of monitored attributes.

4. The specific form of the budget function $B(\gamma)$ is determined through a complex principal-agent problem beyond the scope of this appendix. For insight, see McCubbins and Page (1987), Sappington (1991), and Weingast (1984). In part, the difficulty with analyzing this problem is tied to the mathematical problem of using optimal control theory with constraints that are typically nonconvex. See Rasmussen (1989).

5. An alternative specification could be to assume that $dB/d\gamma$ is some positive constant for output less than the desired quantity, and zero thereafter. For still another concave budget function based on a more explicit treatment of penalties for unacceptable bureaucratic behavior see Bendor, Taylor, and Van Gaalen (1985).

6. The resulting change in regulatory policies suggests rent-avoidance behavior. See Tullock (1980).

7. See Carlson's (1991) description of the U.S. Information Agency's recent experience in producing TV Marti for broadcast to Cuba.

8. For the sake of clarity, we have ignored the benefits to governmental R&D contracts noted in the main body of Chapter 3.

9. See Raviv (1979). Varian (1984) provides a straightforward specification of the demand for insurance for the case of a loss with a fixed probability.

10. The insurance premium $P_1$ will equal $(F_1\text{-}F_c)/2$. Thus, the expected cost to the firm can also be expressed as $F_c+P_1$.

11. See Carlson (1991).

12. See Raviv (1979).

13. The index of absolute risk aversion is defined to be $-U''/U'$ where U is the objective function of the agent in question. It is sometimes known as the Arrow-Pratt measure of risk aversion. See Varian (1984).

14. The privatization of bureaucratic activities is part of a more general issue concerning the appropriate scope of organizations. For an insightful discussion of this more general issue, see Simon (1991) and Stiglitz (1991).

15. See Savas (1987).

# 4

# Investments in Infratechnology Research

As Chapter 3 emphasized, infratechnology is a fundamental factor in explaining the often observed complementarity between governmental R&D allocations and private R&D funding. Buttressed by empirical results, we found that this observed complementarity was the result of technical complementarity at the production level between funding, infratechnology, and knowledge sharing. Thus, investigations of the effect of governmental R&D allocations in stimulating growth in the economy, and therefore the proper role for governmental R&D allocations as part of a rational economic policy, requires that we have accurate measures of the level of infratechnology investment and that we begin to look empirically at the link between such investment and productivity growth in the economy.

Though important to the investigation of productivity growth, the concept of infratechnology is still relatively new to most researchers and policy makers. This chapter presents, for the first time, the results of a systematic effort to collect data on investments in infratechnology research by both Federal laboratories and by industrial manufacturing R&D laboratories,[1] and to report the results of some preliminary investigations of the link between infratechnology investment and productivity growth.

## INVESTMENTS IN INFRATECHNOLOGY

In order to investigate the feasibility of documenting Federal and industrial investments in infratechnology, two populations potentially engaged in infratechnology investment were identified. For Federal investment data, all members of the Federal Laboratory Consortium for Technology Transfer were surveyed. For industrial investment data, a subsample of manufacturing companies from the *Business Week* R&D Scoreboard was surveyed.

### Investments by Federal Laboratories

The 172 Federal laboratories on the October 1990 listing of participating representatives and contacts from the Federal Laboratory Consortium for Technology Transfer were surveyed by mail in 1991. A total of 62 laboratories returned either a partially- or fully-completed survey. This represents a response rate of 36.0 percent.

For this survey (as well as for the industrial survey discussed below) the following definition of infratechnology research was used:

> **INFRATECHNOLOGY RESEARCH** is the process of creating basic scientific and engineering data, measurement and other methods, test procedures, interface definitions, and any other technical entity or procedure which increases the productivity of R&D, production technology, or market transactions for technology-based products.
>
> [An infratechnology acts to **facilitate** the development and use of product and process technologies. Thus, a critically evaluated data base may make R&D more efficient or even possible; a measurement method may be required to improve quality control over a production process; a test procedure may be necessary to prove

compliance with performance specifications; or, a physical or functional interface between two components may be required to allow the components to operate effectively in the same system.]

Of the 62 Federal laboratories returning completed surveys, 29 (46.8 percent) indicated that they were engaged (in 1991) in infratechnology research. As shown in Table 4.1, these laboratories invested in 1990 a total of $1,238.5 million in such research, or, on average, 37.6 percent of their total budget. Their expenditures have more than doubled (in nominal terms) over the last decade, although as a percentage of total budget they have remained fairly constant.

The primary consumer of this infratechnology research is the host laboratory. As shown in Table 4.2, the responding laboratories reported that they were, on average, the most intensive user of their own infratechnology research. In fact, of the 28 laboratories providing this specific information, 19 listed their own laboratory as the primary user.[2] Domestic industries and other Federal laboratories are a distant second and third, with domestic trade associations being the least intensive user of Federally-funded infratechnology research.

## Investments by Industrial Laboratories

A random sample of 417 manufacturing companies, stratified by industry groups, was selected from the 1990 *Business Week* R&D Scoreboard to receive a mail survey in 1991. A total of 126 companies returned either a partially- or fully-completed survey. This represents a 30.2 percent response rate.

The sample of responding companies is summarized in Table 4.3 by two-digit SIC industry groups.[3] Also shown is the ratio of 1989 self-financed R&D by sample of responding companies (as reported in the R&D Scoreboard) to the non-Federal R&D investments in the entire industry.[4] The machinery industry (SIC 35) and the transportation equipment industry (SIC 37) are the more fully represented, having R&D coverage ratios over 20 percent.

Of the 126 companies completing the survey, 57 (45.2 percent) indicated that they were now engaged (in 1991) in infratechnology research. As shown in Table 4.4, these companies (40 of the 57 reported these data) invested $350.9 million in 1990 in infratechnology

**Table 4.1. Federal Laboratory Investment in Infratechnology Research**

| *Year* | *n* | *Total Investment ($M)* | *Mean % of Total Budget* |
|---|---|---|---|
| 1990 | 27 | 1,238.5 | 37.6 |
| 1985 | 25 | 853.3 | 37.1 |
| 1980 | 22 | 615.2 | 38.5 |

research, which represents, on average, 16.3 percent of their total R&D.[5] Infratechnology research expenditures have nearly quadrupled (in nominal terms) over the last decade, and as a percentage of total R&D, they have increased almost 30 percent.

Of the 57 companies engaged in infratechnology research, 47 responded to the survey question regarding infratechnologies as embodied in new capital equipment. On average, 18.4 percent of their 1990 capital budget was for equipment explicitly embodying infratechnologies (e.g., equipment for measurement, testing, process control, data formatting/translating, etc.).

As a final descriptive observation, 86.5 percent of the companies investing in infratechnology research (52 of the 57 responded to this question) did so throughout all of their R&D laboratories in their organization, as opposed to conducting this research within a single dedicated laboratory.

## INFRATECHNOLOGY AND PRODUCTIVITY

As a preliminary step toward understanding the economic growth consequences of these research expenditures, we explored the relationship between total factor productivity growth and manufacturing companies' investments in infratechnology. Specifically, as shown in Table 4.5, data on total factor productivity growth (TFP) between 1979

**Table 4.2. Use of Federal Laboratory Infratechnology Research**

| *Users* | *Mean Score* |
|---|---|
| Your own laboratory | 1.61 |
| Domestic industries | 2.20 |
| Other Federal laboratories | 2.64 |
| Domestic trade association | 3.55 |

*Note:* 1 = most intensive user; 4 = least intensive user.

and 1986 were obtained, by two-digit industry, from the Bureau of Labor Statistics. Also shown in the table are the mean percentages of company R&D expenditures allocated to infratechnology research in 1980 (to allow for a lag), as measured by the survey data from all responding companies (including those not engaged in infratechnology research).[6] These survey-based percentages are intended to approximate each industry's infratechnology research to total R&D ratio.

The Pearson correlation coefficient between these two series is 0.97 (significant at the .01 level). This strong correlation suggests that industries with higher average annual rates of total factor productivity growth are also those that invest a larger portion of their R&D in infratechnology research.

However, care should be exercised when interpreting this finding. First, only five two-digit industry groups were considered. While these are the more R&D-active industries within the manufacturing sector, they certainly do not represent the sector as a whole. Also, it is implicitly assumed in Table 4.5 that the responding companies were a representative cross-section from each industry in terms of overall innovative activity. There is no way to assess the validity of that assumption.

Subject to these caveats, additional correlations were calculated. First, the following cross-industry regression results (n=5) were found to be (t-statistics in parentheses):

**Table 4.3. Survey Sample of Manufacturing Companies**

| *Industry* | *SIC* | *n* | *Coverage Ratio* |
|---|---|---|---|
| Food and kindred products | 20 | 5 | 13.0% |
| Paper and allied products | 26 | 4 | 11.6% |
| Chemicals and allied products | 28 | 23 | 18.7% |
| Petroleum and coal products | 29 | 5 | 5.0% |
| Fabricated metal products | 34 | 4 | 7.9% |
| Machinery, except electrical | 35 | 53 | 21.3% |
| Electric and electronic equipment | 36 | 21 | 8.6% |
| Transportation equipment | 37 | 10 | 27.2% |
| Instruments and related products | 38 | 1 | 0.005% |
| | | 126 | |

*Note:* Coverage Ratio = total R&D in 1989 by sample companies divided by total industry R&D in 1989 (as reported by National Science Foundation).

$$\underset{}{\text{TFP}} = \underset{(-0.01)}{-0.02} + \underset{(1.01)}{0.51}\ \text{RD/S}$$

$$R^2 = 0.25$$

where TFP represents the average industry total factor productivity (from

**Table 4.4. Manufacturing Investments in Infratechnology Research**

| *Year* | *n* | *Total Investment ($M)* | *Mean % of Total R&D* |
|---|---|---|---|
| 1990 | 40 | 350.9 | 16.3% |
| 1989 | 40 | 248.4 | 16.7% |
| 1985 | 29 | 163.5 | 13.7% |
| 1980 | 24 | 92.9 | 12.7% |

Table 4.5) and where RD/S represents the corresponding industry's R&D to sales ratio.[7] This model is similar to the Cobb-Douglas production function model used by others to estimate the rate of return to R&D, as described in Chapter 2. While the magnitude of the coefficient on RD/S is similar to that found by others, it is not significantly different from zero.

Second, an effort was made to estimate the partial effect of investments in infratechnology research on total factor productivity growth. The following cross-industry regression results (n=5) were found to be (t-statistics in parentheses):

$$
\begin{array}{llll}
TFP = & 0.62 & -\ 0.42\ ORD/S & +\ 25.7\ IR/S \\
 & (0.02) & (-1.75) & (5.28)
\end{array}
$$

$$R^2 = 0.95$$

where ORD/S represents the ratio of other R&D to sales and where IR/S represents the ratio of infratechnology-research to sales. Total industry R&D (RD in the numerator of RD/S) equals ORD plus IR.[8] What is especially remarkable about this regression is the statistically significant positive relationship between infratechnology and productivity growth. When combined with empirical evidence that governmental R&D has a positive effect on productivity growth (see Chapter 2) and that

**Table 4.5. TFP Growth and Infratechnology Research**

| *SIC* | *Average TFP (1979-86)* | *Mean % R&D to Infratechnology (1980)* |
|---|---|---|
| 28 | 1.2 | 1.66 |
| 29 | 0.3 | 0.71 |
| 35 | 4.4 | 4.87 |
| 36 | 2.1 | 2.92 |
| 37 | 0.4 | 1.83 |

*Note:* SICs 20, 26, 34, and 38 were eliminated owing to the fact that only one sample company invested in infratechnology research in 1980.

governmental R&D has a positive effect on the level of infratechnology (see Chapter 3), these results persuade us that much of the value of governmental R&D in improving productivity growth comes from its effect on the infratechnology base.

## CONCLUSIONS

The purpose of this chapter was to demonstrate that investments in infratechnology research, an important concept in our theoretical model in Chapter 3, can be quantified in both Federal laboratories and in industrial manufacturing companies. While it is difficult to put the total amount invested by Federal laboratories and manufacturing companies in perspective to their total investments in innovative activity, it is useful to document the intensity with which such research is conducted. On average, Federal laboratories engaged in infratechnology research so allocated 37.6 percent of their total budgets in 1990, and manufacturing companies so allocated 16.3 percent of their R&D budgets.

As a preliminary step toward understanding the economic growth consequences of these research expenditures, the relationship between total factor productivity growth and manufacturing companies' investments in infratechnology was investigated through both descriptive statistics and regression analysis. While care should be exercised due to the many limitations on our data, we find that it is infratechnology research and not R&D *per se* that has a significant and positive effect on total factor productivity. More importantly, these preliminary results (perhaps better called exploratory results) also suggest that to the extent that there is a sharing of infratechnology between private sector firms and governmental laboratories, as documented in Table 4.2, there are notable externalities to productivity growth as Federal laboratories pursue their research agendas.

## NOTES

1. This chapter is based on research funded by the National Science Foundation Science and Engineering Indicators Program. See Link (1991).

2. One laboratory that provided expenditure data did not answer this question.

3. Because *Business Week* uses an alternative industry classification scheme in their R&D Scoreboard, responding companies were re-classified on the basis of their four-digit SIC classification as reported in the Standard and Poor's *Register of Corporations*.

4. The 1989 industry R&D data come from unpublished National Science Foundation tables. The National Science Foundation data are not perfectly matched with the Scoreboard data. The latter could include engineering costs and funding from outside of industry.

5. The ratio of infratechnology R&D to total self-financed R&D was calculated for each company that invested in infratechnology research. The mean of these percentages is reported in Table 4.4.

6. These percentages were calculated by summing investments in infratechnology research by all responding companies, by industry, and dividing by total industry R&D as reported by the National Science Foundation. In 1980, 24 companies (see Table 4.4) reported their actual infratechnology investments. The other 102 companies' investments were $0 for the purpose of calculating these

percentages.

7. These data came from National Science Foundation (1990). Of course, there is no way of knowing if the companies included in the calculation of a specific industry's TFP are grouped in that same industry for the purpose of the National Science Foundation's R&D to sales ratios. The company-financed R&D to sales ratio for 1980 for SIC 37 is unpublished. It was approximated using the average ratio of company-financed R&D-to-sales in SIC 37 to that in SIC 371 for the years 1984 to 1988.

8. The correlation coefficient between ORD/S and IR/S is 0.72, and it is significant at the .20 level.

# 5
# Economic Impact of Federal Investments in Infratechnology

In Chapter 4 we presented preliminary data on the magnitude of public- and private-sector investments in infratechnology research. Albeit that these estimates are preliminary, our evidence suggests that the level of investments in infratechnology research by public- and private-sector participants clearly represents a significant resource commitment to the innovation process. The purpose of this chapter is to extend Chapter 4 by presenting two case studies of investments in infratechnology research by the National Institute of Standards and Technology (NIST; formerly the National Bureau of Standards). Our conclusion is that these investments have a significant impact on the economy -- as suggested in the previous chapters -- and that continued public sector investments appear warranted, at least from a social perspective.

The remainder of this chapter is divided into four parts. First, we present a framework for evaluating the economic impact of these investments. Second, we discuss in detail the two research areas being investigated: NIST's investments to implement standards for optical fiber and NIST's research program on electromigration characterization. Finally, we offer concluding observations on the interpretation of the implications from these two case studies.

## EVALUATING INFRATECHNOLOGY RESEARCH

From an economic perspective, the direct economic value of NIST's (and of most Federal laboratories') infratechnology research can be approximated in terms of the savings associated with reduced "transaction costs" between buyers and sellers, and with increased production efficiency. The associated indirect benefits generally include faster market penetration and increased competitiveness. Such benefits can be seen by examining the two research areas discussed in this chapter -- NIST's research in support of optical fiber standards and NIST's research program on electromigration characterization.

Before NIST's research in optical fiber standards began, buyers and sellers would often have to negotiate at length in order to resolve technical disputes related to the purchase or sale of fiber. These negotiations were typically required when a buyer was unable to verify certain stated performance characteristics of the seller's fiber. Owing to a lack of accepted measurement methods, such negotiations were generally prolonged and expensive, thereby increasing the cost of executing market transactions (hence the term "transaction costs").[1] These costs are just as real as R&D or production costs and were reflected in the effective price paid by the user. Higher prices mean slower market penetration for the technology and, among other things, a greater chance for foreign competition to move in and capture market share. Similarly, prior to the adoption of these standards the production process was longer and less efficient owing to the use of non-uniform testing. Measurement costs were also higher. Thus, this infratechnology research conducted by NIST had the potential to reduce the transaction costs between buyers and sellers of fiber and to increase the efficiency of fiber production.

NIST's research program in electromigration characterization, and its ongoing effort to work with the semiconductor industry to raise its level of measurement expertise and sophistication, also has the potential to reduce the total cost of production. More specifically, these efforts have the potential to reduce the transaction costs between buyers and sellers of semiconductors and to reduce the overall manufacturing costs associated with these devices. Regarding the transaction-costs component of total production costs, it is well known that prior to the entrance of NIST, buyers and sellers would often have to negotiate at length in order to resolve technical disputes related to the reliability characteristics of their chips.[2] In fact, the related standards (discussed below) reduced, and in some cases eliminated, disputes related to such electromigration characterizations.[3] This reduction in disputes means that the costs of executing market transactions has decreased and thus the effective price paid by the user has also decreased. Higher prices mean slower market penetration for the technology and, among other things, a greater chance for foreign competition to capture market shares. A reduction in the overall testing time associated with characterizing the reliability of semiconductors also reduces manufacturing costs. As the manufacturing-costs component of production decreases, there is the potential that the market price will also fall. To the extent that a producer reallocates such cost savings to greater internal testing and or development, product quality may also increase.

Thus, from an economic perspective, the direct economic value of NIST's research programs can be approximated in terms of both the savings associated with reduced transaction costs between buyers and sellers and the savings associated with increased production efficiency and/or reduced manufacturing costs. The indirect benefits are faster market penetration and increased competitiveness.

Figure 5.1 represents the domestic industry supply and the world demand for either optical fiber or semiconductors. Price, P, is measured on the vertical axis and quantity, Q, is measured on the horizontal axis. The domestic industry supply curve is illustrated by the upward sloping supply curve, S. The world market demand (for either fiber or semiconductors) is illustrated by the downward sloping demand curve, D. As illustrated, the market is in equilibrium at price, P*, and quantity, Q*.

Diagrammatically, when these transaction costs and production efficiencies are taken into account, the industry supply schedule will shift to the right to S′. At every quantity, the vertical distance between S′ and S reflects the value to producers of these cost savings. Or stated alternatively, producers now (along S′) have an incentive to supply more at a given price than before they realized these transaction-cost and manufacturing-cost savings. The relevant equilibrium price and quantity are no longer P* and Q*, but rather they are $P^{buyer}$ and Q′ (as determined

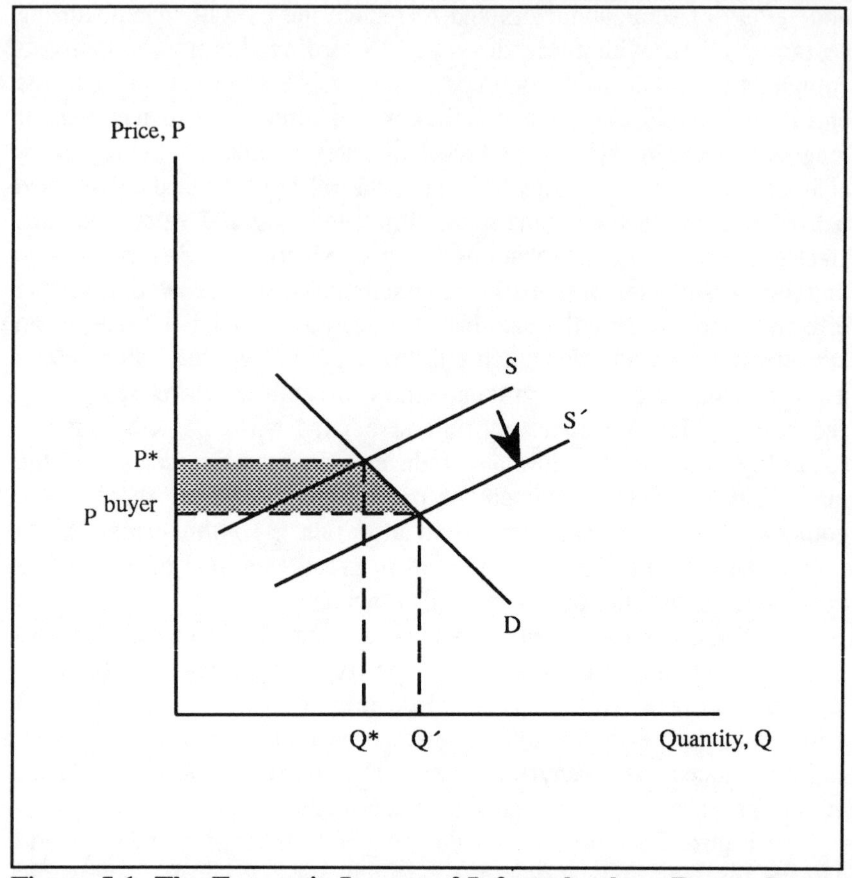


**Figure 5.1. The Economic Impact of Infratechnology Research**

by the intersection of S′ and D). This lower price, $P^{buyer} < P^*$, is brought about by the lower transaction costs and manufacturing costs associated with the production and sale of the product.

The shaded area in Figure 5.1 represents the economic benefits associated with the transaction cost and production efficiency/manufacturing cost savings attributable to the NIST's infratechnology research in either of these areas, or to infratechnology research conducted in other Federal laboratories.

# NIST'S INVESTMENTS TO IMPLEMENT STANDARDS FOR OPTICAL FIBER

## The Optical Fiber Industry

**An Overview of the Technology.** Voice transmission through light beams is not a new technology. As early as 1880 Alexander Graham Bell reported such speech transmission. Many inventors in the early 1900s also experimented with this technology, but they only had limited success. The problems hindering the development and early use of optical communication were two: the lack of a suitable light source and the lack of a suitable transmission medium.[4]

Figure 5.2 shows the basic elements of a simple communication system and a fiberoptic communication system. As seen there, the three critical technologies in the optical fiber system are: (1) a transmitter to convert electrical impulses into light impulses, that is, electrons into photons; (2) a transmission medium (which is the optical fiber cable in the lower part of the diagram); and (3) a receiver to re-code the light impulses into electrical impulses.

Photodiodes (receivers) were widely used in physics-related research in the early 1950s. However, the major technological breakthrough associated with optical communication systems came in the late 1950s, spurred in large part by the research at Bell Laboratories, in the form of the laser (Light Amplification by Simulated Emission of Radiation). With the laser also came the realization that optical communication would be a reality. The only missing element was a suitable transmission medium.

Although the American Optical Company (Massachusetts) and Standard Telecommunications Laboratories (United Kingdom) were early developers of optical fiber, significant technical problems remained. The critical technical barrier to commercialization was described in an academic paper published in 1966 by K.C. Kao and G.A. Hockman which specified the maximum signal loss for commercially-viable long distance transmission (20dB/km, or lower) and "caused [a] stir in the scientific community."[5] The industrial research community knew that the key factor in commercializing optical communication technology would be the fiber core, specifically the development of a glass with minimal absorption of light pulses due to impurities. Corning made the breakthrough in the summer of 1970. The company developed a core

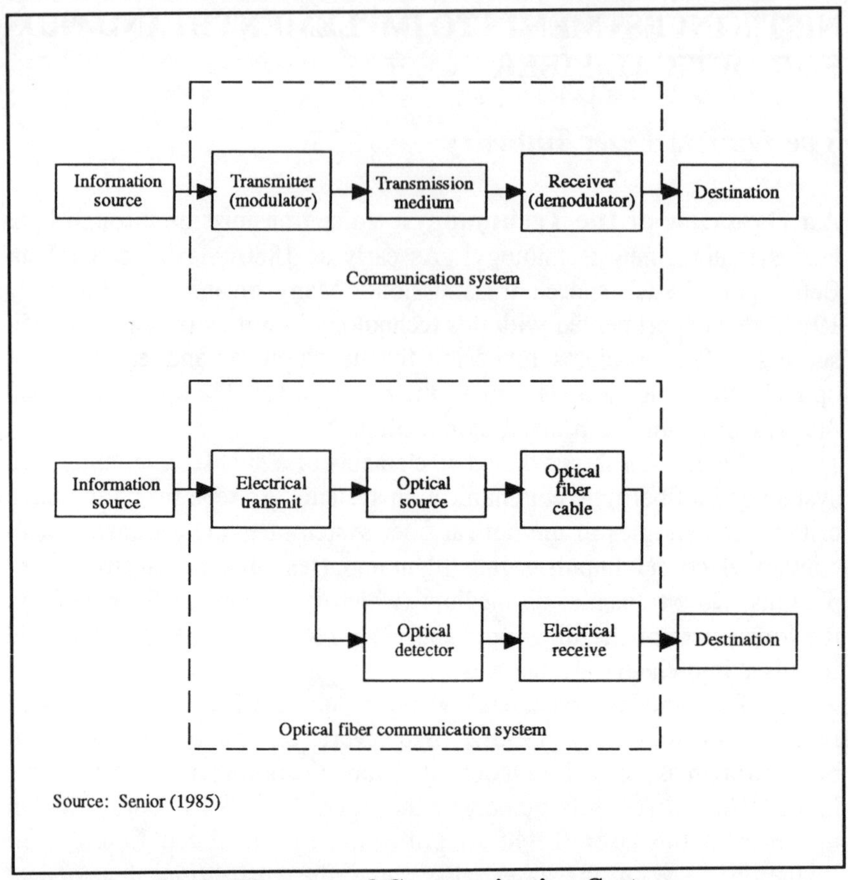


**Figure 5.2. Basic Elements of Communication Systems**

that registered an attenuation rate of 16 dB/km.[6]

Although the technology evolved in the 1970s to the point of allowing commercial communication applications, the market for the technology remained small. Yet, the optical fiber industry maintained its faith in the technology. In fact, Corning began to build its first factory in 1978 (operable in 1979), even before receiving their first sizeable order. AT&T's commitment to the technology was dramatically revealed in 1980 when they announced their intention to build a 611-mile fiberoptic network between Cambridge, Massachusetts, and Washington, D.C.[7]

The 1982 deregulation of the telecommunications industry and

MCI's plan to build an optical communications network to compete with AT&T provided the market with the competitive push it needed. MCI ordered over 100,000 km of fiber from Corning. "The MCI order caught the world off guard -- AT&T, the Japanese, everyone. Corning had the football" and the others would try to take it away.[8] Today, the U.S. optical fiber industry is highly competitive, both domestically and internationally.

Regarding the production process, optical fiber is made in two general stages.[9] Stage I involves the use of one of four vapor deposition techniques to produce silica-rich glass (preforms) from multi-component rods (made from refined glass). Stage II involves the drawing of the preform into fiber.

Stage I was the critical stage in the history of fiber development and, as noted previously, this stage evolved very slowly. All four of the vapor deposition techniques used today are variants of the basic technology developed by Corning in 1972.

Corning's outside vapor deposition (OVD) process, or outside vapor phase oxidation (OVPO) process, consists of building layers of silica soot on the outside of the core rod or mandrel. The layering occurs while the rod rotates over a flame (flame hydrolysis).

Bell Laboratories modified Corning's preform process. Their inside or modified chemical vapor deposition (MCVD) has gas vapor going through a heated silica tube and, as a result, layers of soot accumulate inside.

The Japanese use a vapor axial deposition (VAD) process. Using flame hydrolysis, silica soot is deposited at the bottom of a gathering bar. The rotating preform is then pulled upward to make the fiber.

N.V. Philips (The Netherlands) developed the plasma-activated chemical vapor deposition (PCVD) technique. It is a variation of AT&T's MCVD process, but faster.[10]

The final stage in the production process involves the drawing of the fiber. After it is drawn through a furnace, the fiber passes through a fiber diameter gauge and is then coated with either epoxy or plastic (buffer). Ultraviolet light is used to cure the coating before winding.

There are two general types of optical fiber, multi-mode fiber and single-mode fiber. These two fiber types differ primarily in terms of the size of the core, with multi-mode fiber having the larger one.[11]

Multi-mode fiber was first developed for use with a light-emitting diode (LED). It was thought that LEDs would have a significant cost advantage over lasers. However, as lasers became more reliable and less expensive to manufacture, the use of single-mode fiber increased. Lasers and LEDs are both appropriate as light sources. However, LEDs are less powerful and operate at slower speeds thus making them more suitable for short-hauls and for transmissions with lower information carrying requirements.[12]

Two types of multi-mode fiber are used commercially. Step index fibers have a core diameter of 50 to 400 $\mu$m and are best suited for short-haul, limited bandwidth, low-cost applications. In comparison, graded index fibers are best suited for medium-haul, and medium to high bandwidth applications. These have a core diameter of 30 to 60 $\mu$m.[13]

Single-mode fibers are best suited for long-haul, high bandwidth applications where a single-mode injection laser source is used. The core diameter of these fibers is 3 to 10 $\mu$m. In 1987, shipments of single-mode fiber were more than three times that of multi-mode: 868,000 km compared to 284,000 km.[14]

**The Structure of the Industry.** The optical fiber industry is an oligopoly. There are four domestic producers of fiber, with Corning and AT&T dominating the industry in terms of market share and technological advancements.[15]

Table 5.1 lists these four producers of optical fiber along with their estimated shares of the U.S. market. Together, Corning and AT&T had over 80 percent of this market in the 1983-1986 period, with Corning having the larger share. Industry experts estimate that their combined market share is now about 85 percent. These experts also expect this skewed distribution to remain fairly stable over the next five years, perhaps with Alcatel gaining slightly at the expense of the dominant two.

The value of domestic shipments of U.S. optical fiber was $266 million in 1986 and $275 million in 1989.[16] To place the size of this industry in a broader perspective, the U.S. fiberoptic market (fiber, cable, and components) is about $1.5 billion, while the world fiberoptic market is just over $3 billion and expected to grow to over $10 billion by the year 2000.[17]

Corning also dominates the U.S. market in production capacity. In 1986, its production capacity accounted for over 50 percent of all domestic capacity.[18] See Table 5.2. This inter-firm distribution of production capacities is reported by industry sources to be similar today, with each producer having increased their capacity an average of 25 percent from their 1986 level. However, not all of the capacity is being

**Table 5.1. U.S. Manufacturer Market Shares in the U.S. Optical Fiber Market**

| *Manufacturer* | *Average Market Share (1983-86)* |
|---|---|
| Corning | over 80% |
| AT&T | |
| Alcatel* | 8 - 10% |
| Spectran | 2% |

*Alcatel-Celwave was formerly ITT-Valtec.
*Source:* ITC (1988).

used, or is expected to be used in the early 1990s. As the data in Table 5.3 show, the optical fiber industry had only a 59 percent capacity utilization rate in 1986 compared to an 80 percent rate for U.S. manufacturing industries as a whole. Industry sources expect this percentage to increase to nearly 75 percent by 1995.

U.S. firms (fiber and cable) have relied on similar strategies to acquire/maintain their market shares. As seen in Table 5.4, product quality and price are the dominant strategies. There is no indication from industry experts that this will change in the near future.

**An Overview of Market Trends.** There are two technological issues that will affect the optical fiber industry in the near future. The first relates to Corning's patented technologies and the second relates to newer generations of fiber and related equipment.

It is likely that Corning will continue to dominate the world market owing to their technological leadership in production methods. There is, however, debate about the economic and competitive consequences associated with the expiration of Corning's patents. Some contend that when their current patents expire, Corning will likely introduce new process technologies in order to maintain their

**Table 5.2. U.S. Production Capacity for Optical Fiber**

| *Manufacturer* | *1986 Capacity (km of fiber)* |
|---|---|
| Corning | 1,700,000 |
| AT&T | 1,000,000 |
| Alcatel | 160,000 |
| Spectran | 100,000 |
| Sumitomo Electric* | 70,000 |

*Purchased by AT&T in 1989.
*Source:* ITC (1988).

technological lead.

Current optical fiber technology allows for the spacing of repeaters at about 50 km. Research on new generations of fiber material made from fluoride may increase the distance between repeaters by a factor of possibly 100. If so, transmission costs will fall dramatically, and fiberoptics will become more widely used. Also, new markets will develop in non-telecommunication areas once these new fluoride fibers are commercialized. Better splicing techniques will similarly increase the world demand for optical fibers and fiberoptic technology. Fusion splicing is expected to improve transmission quality by reducing signal attenuation.[19]

Also, it is believed that future cost reductions and quality improvements will come from the integration of the preform and drawing stages into one single continuous production process:[20]

> Japanese and Dutch industry officials and technologists are convinced that their ... processes ... lend themselves well to future continuous process manufacturing systems that will combine the separate steps ... and will result in drastically reduced [production] costs.

**Table 5.3. U.S. Capacity Utilization for Optical Fiber**

| *Year* | *Utilization Rate* |
|---|---|
| 1983 | 100% |
| 1984 | 100% |
| 1985 | 100% |
| 1986 | 59% |
| 1995 *est.* | 75% |

*Source:* ITC (1988) and industry sources.

U.S. officials and industry representatives discount the importance of continuous manufacturing and "do not believe developments in that area will have an impact one way or the other on future global competitiveness in optical fibers."[21]

## Implementing Optical Fiber Standards

Very few technologies can be successfully commercialized and achieve significant market penetration without the availability and use of a critical technology infrastructure. Such infrastructure includes measurement and test methods, data bases used in R&D and process control, and generic models of scientific and engineering phenomena. As with more conventional economic infrastructure, technology infrastructure is used by all market participants more or less equally to leverage productivity and quality and to lower transaction costs. These infratechnologies are often promulgated as standards by industry.

NIST personnel have been very supportive of the optical fiber industry by evaluating, providing technical input, and helping to write relevant standards. The importance of NIST to the optical fiber industry was explicitly documented in a recent report prepared for NIST by Quick, Finan & Associates (1990).[22] They found, based on an

**Table 5.4. Relative Ranking on Alternative Marketing Strategies**

| | |
|---|---|
| 1st: | Product Quality |
| 2nd: | Pricing |
| 3rd: | Technical Service |
| 4th: | Advertising |

*Note:* Rankings in decreasing order of reliance.
*Source:* ITC (1988).

extensive survey of all major U.S. producers of optical fiber as well as several major users, that 60 percent of the firms in that industry had, within the past three years, obtained quality-related information from NIST. These firms reported that they valued highly their association with NIST. In fact, they rated the importance of NIST information at 4.33 on a 5-point scale (from very important (=5) to not important (=1)). The Quick, Finan & Associates report states that these firms explicitly noted the role of NIST in the area of measurement.

The role and impacts of standards in the U.S. economy have been analyzed in a few earlier studies.[23] NIST's role in the standards process, as related to optical fiber, is briefly described below.

During the 1980s, NIST was involved in providing basic measurement technology; evaluating test procedures through interlaboratory comparisons; and offering technical assistance, or helping to write 22 significant fiberoptic test procedures (FOTP) in cooperation with the Electronics Industry Association (EIA). These are listed by EIA's FOTP number and title in Table 5.5.

Table 5.6 categorizes these standards according to the performance-related characteristics of fiber. Eight characteristics are listed in the table: attenuation, cut-off wavelength, mode-field diameter, core diameter, chromatic dispersion, bandwidth, numerical aperture, and geometry. Attenuation is the only one of these characteristics that applies to both single-mode and multi-mode fiber. The other characteristics are fiber-type specific.

**Table 5.5. EIA Fiberoptic Test Procedures Influenced by NIST**

| | |
|---|---|
| **FOTP-29** | Refractive Index Profile Transverse Interference Method<br>(Evaluated in two interlaboratory comparisons, evaluation completed 1982 and 1988; FOTP issued 8/81) |
| **FOTP-30** | Frequency Domain Measurement of Multi-Mode Optical Fiber Information Transmission Capacity<br>(Evaluated internally by NIST, evaluation completed 1981; FOTP issued 9/82) |
| **FOTP-43** | Output Near-Field Radiation Pattern Measurement of Optical Waveguide Fibers<br>(Evaluated in interlaboratory comparison, evaluation completed 1982; FOTP issued 12/84) |
| **FOTP-44** | Refractive Index Profile, Refracted Ray Method<br>(Evaluated in interlaboratory comparisons, evaluations completed 1982, 1988, and 1989; FOTP issued 1/84) |
| **FOTP-46** | Spectral Attenuation Measurement for Long-Length, Graded-Index Optical Fibers<br>(Evaluated in interlaboratory comparisons, evaluation completed 1981; FOTP issued 5/83) |
| **FOTP-47** | Output Far-Field Radiation Pattern Measurement<br>(Evaluated in interlaboratory comparisons, evaluations completed 1981 and 1988; FOTP issued 9/83) |
| **FOTP-50** | Light-Launch Conditions for Long-Length, Graded-Index Optical Fiber Spectral Attenuation Measurements<br>(Evaluated in interlaboratory comparison, evaluation completed 1981; FOTP issued 2/83) |
| **FOTP-51** | Pulse Distortion Measurement of Multi-Mode Glass Optical Fiber Information Transmission Capacity<br>(Evaluated in interlaboratory comparison, evaluation completed 1981; FOTP issued 9/83) |

**Table 5.5. Continued...**

| | |
|---|---|
| **FOTP-54** | Mode Scrambler Launch Requirements for Information Transmission Capacity Measurements (Indirectly evaluated in interlaboratory comparisons, evaluation completed 1981; FOTP issued 9/82) |
| **FOTP-58** | Core Diameter Measurement of Graded-Index Optical Fibers (Evaluated in interlaboratory comparison, evaluation completed 1982; FOTP issued 12/84) |
| **FOTP-78** | Spectral Attenuation Cut-Back Measurement for Single-Mode Fiber by Transmitted Power (Issued 1987; currently under review) |
| **FOTP-80** | Cut-Off Wavelength of Uncabled Single-Mode Fiber by Transmitted Power (Evaluated in interlaboratory comparisons, evaluation completed 1984; FOTP issued 10/88) |
| **FOTP-95** | Absolute Optical Power Test for Optical Fibers and Cables (Evaluated in interlaboratory comparisons, evaluation completed 1988; FOTP issued 6/86) |
| **FOTP-164** | Single-Mode Fiber, Measurement of Mode Field Diameter by Far-Field Scanning (Evaluated in interlaboratory comparisons, evaluations completed 1985 and 1988; FOTP issued 12/86) |
| **FOTP-165** | Single-Mode Fiber, Measurement of Mode Field Diameter by Near-Field Scanning (Evaluated in interlaboratory comparison, evaluation completed 1985; FOTP issued 12/86) |
| **FOTP-166** | Single-Mode Fiber, Measurement of Mode Field Diameter by Transverse Offset (Evaluated in interlaboratory comparison, evaluation completed 1985; FOTP issued 12/86) |

**Table 5.5. Continued...**

| | |
|---|---|
| **FOTP-167** | Mode Field Diameter Measurement by Variable Aperture Method in the Far Field<br>(Evaluated in interlaboratory comparison, evaluation completed 1985 and 1988; FOTP issued 7/87) |
| **FOTP-168** | Chromatic Dispersion Measurement of Multi-Mode Graded-Index and Single-Mode Optical Fibers by Spectral Group Delay Measurement in the Time Domain<br>(Interlaboratory comparison in progress; FOTP issued 7/87) |
| **FOTP-169** | Chromatic Dispersion Measurement of Single-Mode Optical Fibers by Phase Shift Method<br>(Interlaboratory comparison in progress; FOTP issued 8/88) |
| **FOTP-175** | Chromatic Dispersion, Differential Phase Shift<br>(Interlaboratory comparison in progress; FOTP issued 11/89) |
| **FOTP-176** | Measurement Method for Optical Fiber Geometry by Automated Grey-Scale Analysis<br>(Interlaboratory comparison in progress; FOTP not yet issued) |
| **FOTP-177** | Numerical Aperture Measurement of Graded-Index Optical Fibers<br>(Evaluated in interlaboratory comparison, evaluation completed 1988; FOTP not yet issued) |

**Table 5.6. Summary of Single-Mode and Multi-Mode EIA Standards Evaluated or Written by NIST**

| *Characteristic* | *Single-Mode* | *Multi-Mode* |
|---|---|---|
| Attenuation[1] | FOTP 78 | FOTP 46 |
| Cut-Off Wavelength | FOTP 80 | NA |
| Mode-Field Diameter[2] | FOTP 164,167 | NA |
| Core Diameter | NA | FOTP 29,43,44,58 |
| Chromatic Dispersion | FOTP 168,169,175 | FOTP 168,169,175 |
| Bandwidth | NA | FOTP 30,51,54 |
| Numerical Aperture | NA | FOTP 29,44,47,177 |
| Geometry | FOTP 176 | FOTP 176 |

[1]FOTP 50 relates to the light-emitting source launching conditions.
[2]FOTP 164 is used by AT&T, and FOTP 167 is used by Corning.
*Source:* Dr. Douglas Franzen of NIST.

## Estimating the Economic Impacts of NIST-Supported Standards

A two-phase process was undertaken to gather information to approximate the economic benefits associated with NIST-supported optical fiber standards. In early 1990, a survey instrument was designed for obtaining transaction costs information from producers and users of fiber.

The four major producers of optical fiber in the United States are Corning, AT&T, Alcatel, and Spectran.[24] Information from users of fiber was gathered from Bellcore and the U.S. Department of Defense. Bellcore consumes a majority of all private-sector fiber sales, and the Department of Defense is the major public-sector consumer.

After obtaining the survey responses (discussed below), a second more detailed instrument was sent to producers in mid-1991. The purpose of this second phase was to expand on the information collected in the Phase I survey and to collect manufacturing cost information related to the potential impact of standards on this stage in the economic process.

**Phase I Survey Results.** The maintained hypothesis underlying the transaction costs approach to estimating the economic impact of NIST-supported standards is that producers and sellers of fiber will expend fewer resources to resolve technical disputes related to the sale or purchase of the fiber than they would have in the absence of the standards. To verify this maintained hypothesis, each of the four producers was asked to *agree* or *disagree* with the following statement:

> *It is our understanding from industry experts that the economic impact of the measurement standards influenced by NIST has been to reduce the amount of administrative time and associated research activity needed to verify to a potential buyer the stated performance attributes of your fiber.*

All four of the producers surveyed *agreed* with this statement.

Users of optical fiber were asked to *agree* or *disagree* with a similar statement:

> *It is our understanding from industry experts that the economic impact of the measurement standards influenced by NIST has been to reduce the amount of administrative time and associated research activity needed to verify the stated performance attributes of the fiber that you purchase from domestic companies.*

All three of the user groups of fiber surveyed *agreed* with this statement.

Each producer and user was also asked to indicate the performance attributes affected by measurement standards (see the characteristics listed in Table 5.6) from which they realized transaction cost savings. The respondents checked on the surveys all of the characteristics, thus indicating that the transaction costs aspect of the economic impact of any one standard could not be accurately determined.[25]

For the group of optical fiber producers (in other words, the U.S. domestic industry), their total current (1990) estimated annual transaction-cost savings from NIST-sponsored standards was $2.8 million.[26] For the group of users, their total current (1990) estimated annual transaction-cost savings was $13.5 million.[27] Although there was no *a priori* estimate of the magnitude of the difference in transaction cost savings between producers and users, it was expected that the greater savings would occur to users owing to, among other things, their relative lack of technical measurement expertise.[28] The survey estimates show that these savings to users are about five times those to producers.

A reasonable first-order estimate of the economic impact of NIST-sponsored optical fiber measurement standards in 1990 was $16 million, the sum of the average reported transaction-cost savings estimates to producers and users of fiber.

Each respondent was also asked:

> *In the absence of NIST, do you think these standards [listed in Table 5.5] would have been developed by industry participants?*

Three of the four producers answered this question, *Yes*. All three of the users answered *Yes*, too. There was unanimous agreement among the six that alternative standards would have been developed, but with considerable delay. And, they all believed that NIST's participation in the process led to superior standards being developed than would otherwise have been the case. In fact, follow-up telephone interviews with a sample of respondents suggested that the standards listed in Table 5.5 may have in fact taken two to three times as long to develop in the absence of NIST.[29]

The objective information from the Phase I surveys from producers and users of optical fiber can be used to approximate an annual time series of transaction cost savings data. We assumed that all 22 of the NIST-supported standards listed in Table 5.5 contributed equally to the 1990 estimated savings of $16 million. Then, based on the respondent's estimate that these standards would have taken two to three times as long to develop in the absence of NIST, only between 50 and 67 percent of these savings can be attributable directly to NIST. Thus, in 1990, NIST research in measurement-related optical fiber standards accounted for between $8 million and $11 million of cost savings, or an average of $9.5 million.[30]

NIST's formal involvement in the optical fiber measurement area began in 1981. An inspection of the EIA fiberoptic test procedures (FOTPs) listed in Table 5.5 shows that several evaluations were completed in 1981 and some FOTPs were issued in 1982. It seemed reasonable to us, after discussions with NIST and industry experts, to expect that some benefits were realized in 1982, and that these benefits increased thereafter. If growth was linear from 1982 through 1990, then Table 5.7 shows the transaction-cost savings to producers and users over time. Therefore, based on these estimates, the cumulative transaction-cost savings (economic benefits) from NIST's standards research has been, through 1990, just over $47 million.

**Phase II Survey Results.** Three of the four producers of optical fiber were willing to participate in the second phase of data collection: AT&T, Corning, and Alcatel. The savings estimates in Table 5.7 were constructed, in part, on the assumption that the 22 NIST-supported standards listed in Table 5.5 contribute equally to transaction-cost savings. Two of the three producers answered *Yes* to the question:

> *Is it possible to attribute [transaction-cost savings] to a particular set of standards ... ?*

However, these two companies did not agree on which standards should be in this group. One company defined the group to include 13 of the 22 FOTPs in Table 5.5 and the other defined it to include only four. Perhaps this difference is attributable, in part, to the fiber mix of their production processes. For the analysis that follows, no effort is made to allocate the cost-savings estimates by type of standard.

During the Phase I follow-up telephone interviews, one major producer noted that his company realized benefits in addition to those associated with reduced transaction costs. He reported that NIST-supported standards have increased his company's sales by a factor of about 25, of which 80 percent was domestic. No specific additional data on this issue could be collected for the group of companies in the Phase II survey.

The conceptual model of economic impact presented above considered both transaction cost savings and manufacturing cost savings as elements of the economic benefits associated with NIST-supported

**Table 5.7. Transaction-Cost Savings Attributable to NIST-Sponsored Research**

| *Year* | *Savings ($million)* |
|---|---|
| 1981 | 0.00 |
| 1982 | 1.05 |
| 1983 | 2.11 |
| 1984 | 3.16 |
| 1985 | 4.21 |
| 1986 | 5.26 |
| 1987 | 6.31 |
| 1988 | 7.36 |
| 1989 | 8.41 |
| 1990 | 9.50 |
| | $47.37 |

*Note:* Savings were defined to be $1.05 \times T$ where $T=0$ in 1981.

standards. As a result, we asked:

> *In addition to transaction-cost savings, have [the NIST-sponsored] standards increased the efficiency of your production process?*

Two producers responded *No*, and one responded *Yes*. In fact, one of the producers who responded *No* stated, "these standards do not save [production] costs, they obviously increase costs but because of the standards and the confidence ... they engender, the size of the business and consequently of sales has increased at a very high rate." The one producer who responded *Yes* estimated average cost savings from production efficiency gains to be between $50,000 and $150,000 per year for his company.

Therefore, the transaction-cost savings estimates in Table 5.7 are a reasonable lower-bound estimate of the direct economic benefits to optical fiber producers and users from NIST's research program in measurement-related optical fiber standards. Although the Phase II survey suggested that there are second-order benefits in terms of increased sales and production efficiencies, no industry-wide dollar estimates could be obtained.

Finally, we thought it was reasonable to ask:

> *If NIST eliminated its research program in measurement-related optical fiber standards, approximately how many years into the future would your company continue to receive cost-related benefits?*

The mean estimate from the three responses was 5 years. In addition, the three producers responded that all 22 of the NIST-sponsored standards are still relevant. Two producers thought the benefits from this group of standards would decline moderately over the next five years and one producer thought the decline would be sharp.[31]

Therefore, a lower-bound estimate of the total economic benefits from NIST's research program in measurement-related optical fiber standards are those values in Table 5.7, plus the expected cost-savings benefits over the next five years. Shown in Table 5.8 is one possible extrapolation of benefits into the future. This particular extrapolation declines uniformly.[32] Past and future benefits from NIST's research program are estimated to be just over $65 million.

## Implications of the Findings

As shown in Table 5.8 and discussed above, the total economic benefits attributable to NIST's research program in measurement-related optical fiber standards are at least $65 million. There is reason to believe that this is a conservative estimate owing to the fact that the economic value of both increased sales and improved production efficiency were not included.

Society as a whole also benefits from NIST's activity to the

**Table 5.8. Past and Future Transaction-Cost Savings Attributable to NIST-Sponsored Research**

| *Year* | *Savings ($million)* |
|---|---|
| 1981 | 0.00 |
| 1982 | 1.05 |
| 1983 | 2.11 |
| 1984 | 3.16 |
| 1985 | 4.21 |
| 1986 | 5.26 |
| 1987 | 6.31 |
| 1988 | 7.36 |
| 1989 | 8.41 |
| 1990 | 9.50 |
| 1991 | 7.50 |
| 1992 | 5.50 |
| 1993 | 3.50 |
| 1994 | 1.50 |
| 1995 | 0.00 |
| | $65.37 |

extent that producer and user cost savings lead to lower market prices, as was shown in Figure 5.1, and production efficiencies lead to a higher quality fiber which in turn benefits society first through improved fiberoptic technology and second through a more competitive domestic industry in the global optical fiber market. Therefore, the cost-savings estimates in Table 5.8 should be viewed as a lower-bound estimate of the social benefits of NIST research.

These social benefits were not achieved without a social cost. As shown in Table 5.9, the research costs to generate the economic benefits have increased each year since 1981. These research costs are divided into two categories. The first category includes only direct NIST research costs. However, other Federal agencies did (and still do) contribute to NIST's optical fiber research program. These other agency (OA) costs are included in the second cost column labeled NIST+OA. NIST's financial contribution to the total optical fiber research program budget has increased from 60 percent in 1981 to 83 percent in 1989.[33] Through 1989, direct NIST support totaled about $2.8 million.

The benefits to society from these public expenditures, that is the social rate of return to these research dollars, can be approximated by the rate of return, r, that satisfies the following equation:

$$\sum_{i=0}^{t} B_i/(1+r)^i = 0 \qquad (5.1)$$

where $B_i$ is the net social benefit in the ith year after NIST's research program in measurement-related optical fiber standards began.[34] An empirical estimate of r represents the rate of return society earned on such investments.

Annual net benefits, $B_i$, were calculated as the difference between annual social benefits (Table 5.8) and annual social costs (Table 5.9). These net benefits are in Table 5.10 for both direct NIST costs and NIST+OA costs. Based on the net benefits associated with NIST+OA costs, the social rate of return to NIST research in measurement-related optical fiber standards is 423 percent.

**Table 5.9. Costs Associated with NIST-Sponsored Research**

| *Year* | *Cost (in $million) of ...* | |
|---|---|---|
| | *NIST* | *NIST + OA* |
| 1981 | 0.150 | 0.248 |
| 1982 | 0.150 | 0.228 |
| 1983 | 0.192 | 0.336 |
| 1984 | 0.310 | 0.396 |
| 1985 | 0.228 | 0.376 |
| 1986 | 0.354 | 0.491 |
| 1987 | 0.397 | 0.537 |
| 1988 | 0.495 | 0.550 |
| 1989 | 0.505 | 0.610 |
| | $2.781 | $3.772 |

# NIST'S RESEARCH PROGRAM ON ELECTROMIGRATION CHARACTERIZATION

## Electromigration Characterization

Electromigration is a failure mechanism of the thin-film metal interconnect lines which serve as the "wiring" to connect electrically the various components of integrated circuits. The width and thickness of these metal lines can be as small as a micrometer or less. Hence, while the current through these "wires" may be measured in milliamperes, the

**Table 5.10. Net Social Benefits from NIST's Research Program**

| *Year* | *Net Benefits (in $million) from ...* | |
|---|---|---|
| | *NIST* | *NIST + OA* |
| 1981 | -0.150 | -0.248 |
| 1982 | 0.900 | 0.822 |
| 1983 | 1.918 | 1.774 |
| 1984 | 2.850 | 2.764 |
| 1985 | 3.982 | 3.834 |
| 1986 | 4.906 | 4.769 |
| 1987 | 5.913 | 5.773 |
| 1988 | 6.865 | 6.810 |
| 1989 | 7.905 | 7.800 |
| 1990 | 9.500 | 9.500 |
| 1991 | 7.500 | 7.500 |
| 1992 | 5.500 | 5.500 |
| 1993 | 3.500 | 3.500 |
| 1994 | 1.500 | 1.500 |
| 1995 | 0.000 | 0.000 |
| | $62.589 | $61.598 |

current density (current divided by "wire" area) can be almost a million amperes per square centimeter. This contrasts with a current density of 600 amperes per square centimeter in a 12 AWG (American Wire Gauge) gauge wire, used in most homes, at its rated current of 20 amperes. As current densities approach a million amperes per square centimeter in thin-film metal lines, the electrons (which constitute the current) will cause the metal atoms to move along the grain boundaries of the metal film in ways that can create accumulations of the metal in some regions and voids in the metal film in other places of the conducting line. Given enough time, failure of the line can occur due to either an accumulation of metal at some point in the line, causing a short circuit to an adjacent metal line, or to a coalescence of voids, causing an open circuit. The ability of a metal line to sustain high current densities without failure due to electromigration depends, in ways that are not well understood, on many factors, including: the metal and metal alloy, parameter control of processing materials and gases, the metal-deposition process and its control, and subsequent processing.

Electromigration has been a continuing concern to those involved in the reliability assessment of VLSI-sized[35] microelectronic devices. An important aspect of this concern has been the ambiguities in the electromigration characterization of metallizations due to the different test structures and measurement methods used and due to the incomplete reporting of results from these characterizations.[36]

The inability to characterize with confidence the resistance of metallization to this failure has frustrated the industry. In fact, at the 1983 Wafer Reliability Assessment Workshop, sponsored by DARPA (Defense Advanced Research Projects Agency), U.S. semiconductor representatives expressed grave concern about the need to improve both the quality and the reproducibility of measurement results for evaluating and comparing the electromigration susceptibility of metallizations.[37] NIST, then the National Bureau of Standards, was specifically called upon to assist the industry. NIST undertook a leadership role to conduct an "interlaboratory experiment with the industry which [was] intended to promote the reproducibility of electromigration characteristics. [NIST was] also working with the industry and universities to develop guidelines and standards for electromigration testing and reporting."[38,39]

NIST's continued involvement with the industry has been through participation in the Fine Line Conductor Task Force meetings sponsored by RADC (Rome Air Force Development Center, now Rome

Laboratory), the DARPA-sponsored Wafer Level Reliability Assessment Workshops, the International Reliability Physics Symposia, ASTM and JEDEC standards committees, and collaborative work with companies within the semiconductor industry and with universities.

The outputs from this participation include:

- *Interlaboratory Electromigration Experiment*: to assess the industry's measurement capability for making electromigration characterizations and to identify sources of measurement error.[40]

- *Thermal Analysis of Electromigration Test Structures*: to develop a basis for designs to minimize temperature and other gradient interferences to measurement precision.[41]

- *Identification of Thermal Interactions*: to account for thermal interferences to measurement precision.[42]

- *Statistical Analysis Procedures*: to provide a basis for analyzing electromigration test data.[43]

- *ASTM Electromigration Standards*:[44]

  F-1259-89: Standard Guide for Design of Flat, Straight-Line Test Structures for Detecting Metallization Open-Circuit or Resistance-Increase Failure Due to Electromigration

  F-1260-89: Standard Test Method for Estimating Electromigration Median Time-to-Failure and Sigma of Integrated Circuit Metallizations

  F-1261-89: Standard Test Method for Determining the Average Width and Cross-Sectional Area of a Straight, Thin-Film Metal Line.

- *Pulsed Electromigration Studies*: to identify measurement interferences in making pulsed electromigration characterizations and relating them to dc conditions.[45]

- *JEDEC Standard for Temperature Coefficient Resistance*: to respond to industry's request (under development in Committee JC 14.2).

## The U.S. Semiconductor Industry

**Industry Overview.** Since the invention of the transistor in the early 1950s, semiconductors have become an integral part of most industrialized economies. In fact, worldwide sales of semiconductors have increased from $40 million in 1955 to $5 billion in 1975, to $29 billion in 1985, to $50 billion in 1989.[46] Worldwide sales are forecast to surpass $200 billion before the end of the century.[47]

Yet, even this magnitude of sales of the world semiconductor industry understates its economic importance. "Semiconductor technology is unique in being not only the foundation of a large industry ... but the key technology required for making all electronic systems."[48] "Today's $50 billion world chip industry leverages a $750-billion dollar global market in electronics products and 2.6 million jobs in the United States."[49] This represents more than double the number of jobs in the U.S. steel and automobile industries combined. "Over the past three decades, no single field of science or engineering has had a greater impact on the national productivity and quality of life in the United States than has semiconductor microelectronics."[50] Not only is American industry dependent on advances in semiconductor technology, but also America's national security rests on its ability to maintain a technological advantage in this area.[51]

However, numerous indicators suggest that the United States is losing its competitive advantage in semiconductor technology. For example, the U.S. share of the world merchant semiconductor industry has declined steadily since the early 1970s, as shown in Table 5.11.[52] Perhaps of greater importance is the decline in the U.S. share of the world market for memory devices. "Memory devices are the key component for computers, consumer electronics, telecommunications, industrial electronics, and defense electronics. They ... serve as a ... 'technology driver,' pushing a broader range of semiconductor device technology to the limits of perfecting manufacturing processes."[53]

**Table 5.11. Merchant Semiconductor Industry Market Share History**

| *Year* | *U.S. %* | *Japan %* | *Europe %* | *Rest of World %* |
|---|---|---|---|---|
| 1952 | 100.0 | | | |
| 1955 | 95.0 | | 5.0 | |
| 1960 | 78.0 | 10.0 | 11.0 | |
| 1965 | 73.0 | 12.6 | 14.3 | |
| 1970 | 56.5 | 27.1 | 16.1 | |
| 1975 | 63.9 | 19.3 | 16.7 | 0.1 |
| 1980 | 42.9 | 24.0 | 26.1 | 7.0 |
| 1985 | 38.7 | 33.5 | 19.9 | 8.0 |
| 1990 | 31.7 | 38.4 | 17.4 | 12.5 |
| 1994(est.) | 30.8 | 37.2 | 17.7 | 14.4 |
| 1999(est.) | 29.3 | 35.6 | 17.3 | 17.8 |

*Note:* Data include intracompany consumption for merchant suppliers and exclude consumption for full captives.
*Source:* Scace (1990).

During the 1970s, the United States controlled 100 percent of the world market for dynamic random access memory devices (DRAMs). By 1982, Japan's share surpassed that of the United States. In 1989, Japan had nearly 80 percent of this vital market.[54]

**Semiconductor Technology.**[55] The progression of the technology for making semiconductor devices has gone through several stages. In the early stages, silicon transistors and diodes were fabricated and used separately as replacements for most vacuum tubes in electronics equipment. Later came the development of the concept of fabricating many of these individual devices together and having them electrically connected to operate as a circuit. This led, at first, to the fabrication of simple, integrated circuits consisting of several of these devices. Thin-film metallization was used to connect electrically these devices.

Over the years, the technology has advanced to the manufacture of today's integrated circuits (ICs) that contain as many as 10 million transistors. ICs containing well over 200 million transistors are in development. In this progression, the technology has led to shrinking dimensions and ever greater circuit density and circuit performance. To effect the electrical connection of these many circuit components, more than one level of these thin-film metallizations have been used which, in combination, cover most of the area of the circuit. These many thin-film metallization interconnects must function reliably. Where they have not, the expenses that have been incurred by the device manufacturer have been measured in many millions of dollars.

The manufacture of these ICs is a complex sequential process of up to approximately 500 steps. The process of making semiconductor devices starts with very flat, round wafers of silicon which are obtained from the material supplier who has cut them from ingots of single-crystal silicon that are about a meter long. The wafers are typically 100 to 200 millimeters in diameter and polished by the supplier to be flat to within less than 3 micrometers. In addition to the rigid requirements on the flatness of these wafers are the requirements on the crystallographic perfection and chemical purity.

The wafers are processed in special facilities in which the temperature, humidity, and particulate contaminations are tightly controlled. Extreme attention to cleanliness is necessary to avoid chemical and particulate contamination of the silicon wafers during handling.

The processing of the wafer involves a series of steps involving the deposition or formation of various films in or on the wafer surface. A part of these steps is the selective removal of these films, creating patterns of these films on the wafer which function as some part of the devices being built. This processing builds a structure of perhaps 20 different layers, all interconnected laterally and vertically to form a complex network of individual transistors, diodes, resistors, and capacitors, and the electrical connections among them.

Films may be dielectrics such as silicon dioxide, or conducting materials such as heavily-doped silicon or metals. The films are deposited by evaporation or sputtering in a vacuum, or by chemical processes involving reactions promoted in a mixture of gases by heat or electrical energy in a partial vacuum. The quality and reliability of the metallization and the other films depend critically on a multitude of circuit design features, material characteristics, processing variables, and subsequent processing in complicated ways that are still not well characterized or understood.

**Technological Trends.** The hallmark of the evolution of electronic and semiconductor devices has been the reduction in their scale of dimension. This trend is clearly illustrated in terms of the evolution of the electronic devices in Table 5.12. The individual elements are becoming smaller so that the circuits can become more complex and thus more powerful. Regarding memory devices such as DRAMs, memory capacity was increased from 1kb (1000 bytes[56] or 1 kilobyte) in commercialized circuits in 1970, to 4kb in 1974, to 16kb in 1976, to 256kb in 1982, to 1Mb (1 million bytes or 1 megabyte) in 1985, to the experimental 4Mb in 1987 with commercialization in 1989, and to the 1Gb (1 billion bytes or 1 gigabyte) DRAM by the year 2000.[57]

Manufacturing a semiconductor device is a measurement-intensive activity. The importance of measurement will even increase over time as elements become smaller and devices become more complex. It has been estimated that measurement activity now represents 25 percent of the cost of producing a semiconductor device.[58]

The areas where measurement technology is most important are product development and design, materials procurement, wafer fabrication, and assembly/testing.[59] Semiconductor manufacturers obtain measurement technology from several sources. The most important are their own R&D divisions, NIST, standards bodies, and trade associations.[60] In general, as measurement becomes more important in production, interactions with these external organizations will also become more important.

## Estimating the Economic Impact of Electromigration Research

In order to quantify the direct economic impact (i.e., cost savings) of NIST's research program and its work with the

**Table 5.12. Evolution of Electronic Devices**

| *Dimension* | *Technology* | *Relevant Years* |
|---|---|---|
| Centimeter (cm) | Vacuum Tube | 1925-1960 |
| Millimeter (mm) | Transistor | 1950s- |
| Micrometer ($\mu$m) | Integrated Circuit | 1980s- |
| Sub-Micrometer | Advanced Integrated Circuit | 1990s |

*Source:* Scace (1990).

semiconductor industry, a survey of domestic producers was conducted in late 1991.

Contact individuals at AT&T, Hewlett-Packard, IBM, Intel, Motorola, National Semiconductors, SGS-Thomson Microelectronics, Texas Instruments, VLSI Technology, and Xicor were identified by the Electronics and Electrical Engineering Laboratory (EEEL) at NIST. This sample includes six of the seven largest U.S. semiconductor companies. Each was sent a mail survey to complete. Eight of the ten companies responded to the survey. Follow-up telephone interviews were conducted with all ten companies.

To understand the indirect economic impact of NIST's research in electromigration characterization, two other surveys were conducted. One survey went to academic researchers in this area and the other went to testing systems makers.

Regarding academic researchers in the area of electromigration, five of the leading researchers were identified with the assistance of EEEL.[61] Each was sent a mail survey. One individual responded; however, the other four answered selected questions during follow-up telephone interviews.

Regarding the testing systems makers, EEEL identified three: Micro Instruments Co., Reedholm Instruments, and Sienna Technologies. One company's representative responded by mail and the two other representatives were interviewed by telephone. These companies are the major testing systems makers that market electromigration test stations.

**Direct Economic Impact.** All eight of the semiconductor manufacturing companies' respondents agreed with the following survey statements:

> *It is our understanding that the overall economic impact of NIST's research program in electromigration characterization ..., has been to give producers a greater degree of confidence in characterizing their metallization.*
>
> *It is our understanding that one important result of this has been to reduce the time (and hence cost) incurred in resolving measurement discrepancies between vendors and users.*
>
> *It is also our understanding that another important result has been to increase the efficiency (reduction in time, increased ability to interpret results) of your R&D related to (1) the evaluation of alternative metal alloys or (2) evaluation or optimization of alternative deposition processes.*

When asked to estimate the approximate cost savings (time saved, improved R&D efficiency, production cost reduction, etc.) to the company from implementing the results of NIST's research program in electromigration, only five responded. The other three noted difficulty in answering such a question and left it blank. Accordingly, all ten of the companies were interviewed by telephone to validate the accuracy of the quantitative information reported on this question and to elaborate upon this question in an effort to obtain a lower-bound estimate from the other five companies. Quantitative estimates were obtained from the missing five companies; the five companies originally reporting cost-savings estimates verified the accuracy of their initial responses. For 1991, the estimated total cost savings to the U.S. semiconductor industry from NIST's research program (as proxied by the responses from these ten companies) was $4.1 million.[62]

During the telephone interviews with each company, quantitative information was also collected on past and expected future cost savings. Most of the companies noted that they began to realize cost-savings benefits around 1988, just prior to the formal adoption of the three ASTM electromigration standards referenced above. Two companies said that they began to receive some cost-savings benefits as early as 1984. In addition, each company was asked to estimate how long into the future they would continue to realize cost-savings benefits if NIST eliminated its research program in electromigration and ceased to assist industry with related-measurement issues. Estimates ranged from one to two years, up to ten years.

Table 5.13 shows the total industry benefits (cost savings) from NIST's research program in electromigration, by year, as reported by the these ten companies. This benefit stream is noticeably larger in the 1988 to 1992 period. The total reported benefits, past and expected future as reported by the survey respondents, are over $26.6 million from NIST's research program through 1991.

This $26.6 million cost savings represents the direct economic impact of NIST's activities on the semiconductor industry. There is reason to believe that this estimate is conservative for at least two reasons. First, not all producers in the semiconductor industry were surveyed. It may well be the case that smaller producers have benefited disproportionately to their size from NIST's involvement with the industry. Second, the estimates presented in Table 5.13 are lower-bound estimates. A number of companies reported in the telephone interviews that their benefits were "at least $X, but likely more." Only their $X are reported in Table 5.13.

**Indirect Economic Impact.** The survey responses from the academic researchers and the testing systems makers clearly indicate that there are indirect benefits from NIST's research in electromigration. For example, the total-to-date academic-related cost savings could be approximated at $11.7 million based on the average cumulative cost savings from the five academics and from their average estimate of the number of similar researchers. However, because no information was available on the representativeness of this sample of five, this dollar estimate was not included in the social rate of return estimate below.

No cost-savings estimates could be obtained from the testing systems makers. However, each respondent did suggest that the industry would not even exist were it not for NIST.

**Table 5.13. Semiconductor Industry's Benefits from NIST's Electromigration Research Program**

| *Year* | *Annual Benefits ($thousands)* |
|---|---|
| 1981 | 0 |
| 1982 | 0 |
| 1983 | 0 |
| 1984 | 300 |
| 1985 | 300 |
| 1986 | 300 |
| 1987 | 575 |
| 1988 | 3,775 |
| 1989 | 3,975 |
| 1990 | 4,025 |
| 1991 | 4,125 |
| 1992 | 4,175 |
| 1993 | 1,600 |
| 1994 | 1,375 |
| 1995 | 935 |
| 1996 | 700 |
| 1997 | 100 |
| 1998 | 100 |
| 1999 | 100 |
| 2000 | 100 |
| 2001 | 100 |
| | $26,660 |

**Table 5.14. NIST's Annual Costs for Activities Related to Electromigration Through 1991**

| *Year* | *Annual Cost ($thousands)* |
|---|---|
| 1981 | $ 10.5 |
| 1982 | 55.4 |
| 1983 | 122.6 |
| 1984 | 124.0 |
| 1985 | 121.9 |
| 1986 | 130.1 |
| 1987 | 151.6 |
| 1988 | 195.5 |
| 1989 | 241.1 |
| 1990 | 258.7 |
| 1991 | 242.2 |
| | $1,653.6 |

## Implications of the Findings

Society as a whole benefits from NIST's research activity in electromigration beyond both the direct benefits to the semiconductor industry and the indirect benefits to academicians and testing systems makers. Producer's cost savings also lead to a higher quality semiconductor product that is sold at a lower price, and university-based research enriches society's knowledge base.

The benefits to society from these public expenditures, that is the social rate of return to these research dollars, can be approximated by the rate of return (or discount rate) that equates the present value of all

**Table 5.15. Net Social Benefit from NIST's Electromigration Research Program**

| *Year* | *Net Benefits ($thousands)* |
|---|---|
| 1981 | -10.5 |
| 1982 | -55.5 |
| 1983 | -122.6 |
| 1984 | 176.0 |
| 1985 | 178.1 |
| 1986 | 169.9 |
| 1987 | 423.4 |
| 1988 | 3,579.5 |
| 1989 | 3,733.9 |
| 1990 | 3,766.3 |
| 1991 | 3,882.8 |
| 1992 | 4,175.0 |
| 1993 | 1,600.0 |
| 1994 | 1,375.0 |
| 1995 | 935.0 |
| 1996 | 700.0 |
| 1997 | 100.0 |
| 1998 | 100.0 |
| 1999 | 100.0 |
| 2000 | 100.0 |
| 2001 | 100.0 |
| | $25,006.4 |

future benefits with the present value of all past and future costs using equation (5.1).

Here, a quantitative estimate of r represents the rate of return society earned from the cost savings received by the U.S. semiconductor industry from NIST's research activities in electromigration.

Annual net benefits, $B_i$, were calculated as the difference between annual industry benefits (Table 5.13) and annual NIST costs (Table 5.14). These estimates are shown in Table 5.15. Using these estimates, the value of r that satisfies the condition in equation (5.1) was estimated to be 117 percent.

## CONCLUSIONS

The findings from these two case studies substantiate the benefits associated with NIST's investments in infratechnology research. The estimated social rates of return of 423 percent (optical fiber) and 117 percent (electromigration characterization) compare favorably with estimated social rates of return in other studies.[63]

While these two areas of infratechnology research were selected, in part, on the basis of NIST's interest, we have no reason to believe that similar infratechnology research at NIST and at other Federal laboratories does not make the same broad-based contribution to industry as do these programs. In fact, these findings reinforce our belief that the Nation's technology infrastructure is perhaps the critical factor that will determine the long-run competitive position of U.S. industry in the global economy.

## NOTES

1. The transaction costs that producers and buyers incurred during negotiations in order to resolve technical disputes related to any one of the fiber's performance characteristics involve monetary expenditures. These may include verification costs, documentation costs, or simply the added time devoted to the negotiation process. Similarly, non-uniform testing and duplicated measurements decrease production efficiency and thus add to the overall cost of manufacturing.

2. One way to characterize reliability is by median time-to-failure, or $t_{50}$. It has been conjectured that prior to the formulation of F-1260-89 (see below), one company's determination of $t_{50}$ would differ from those of one of their important customers by a factor of 50. See Schafft (1989).

3. Many companies actively participated in critiquing draft standards. In fact, the draft standards were widely used before the final standards were accepted.

4. As Senior (1985) notes, these early experimenters found transmission through the atmosphere to be plagued with too many sources of interference.

5. See Chaffee (1988) for more detail and for the original reference of the Kao and Hockman paper.

6. See Magaziner and Patinkin (1989) for a more detailed history of Corning's research in this area.

7. See Chaffee (1988).

8. See Magaziner and Patinkin (1989), p. 290.

9. This section draws from ITC (1988) and Senior (1985).

10. See ITC (1988).

11. In addition, there is plastic-clad fiber and all-plastic fiber. Both have relative large core diameters. They are less expensive than glass fiber, but they have limited performance characteristics and thus limited communication applications. Dupont dominated the U.S. market for plastic fiber, but discontinued production in 1982. They sold their technology and production rights to Mitsubishi Rayon, which now has about 90 percent of the world market for plastic fiber. In 1987, Dow Chemical announced its intention to enter this market; however, industry sources say that this is unlikely to occur. See ITC (1988).

12. See ITC (1988).

13. 50 $\mu$m has been established as the standard for telecommunication applications.

14. See ITC (1988).

15. AT&T purchased Sumitomo Electric in 1989.

16. This is based on U.S. Department of Commerce data for SIC 32318.

17. See ITA (1988).

18. This is the latest published information.

19. The use of optical fiber for telecommunications has increased steadily in recent years. Whereas only 10 percent of total telecommunication route miles was optical fiber in 1984, over 40 percent was optical fiber by the end of 1990. This increase has come at the expense of a decrease in the use of microwaves, copper wire, and satellites. The use of coaxial cable has remained constant. See ITC (1988).

20. See ITC (1988), p. 3-7.

21. See ITC (1988), p. 3-7.

22. See also Link, Quick, and Tassey (1991).

23. See Link and Tassey (1987).

24. LITESPEC was formed in February, 1989. It is a joint venture between Sumitomo Electric Industries of Japan and AT&T. Formerly, Sumitomo's fiber was produced in the United States by Sumitomo Electric.

25. During the pre-testing of the Phase I instruments, industry experts suggested that it would not be possible to attribute transaction-cost savings from NIST-supported standards to either a sub-group of standards or to one specific standard. This *a priori* belief was substantiated.

26. The range of responses from the individual producers was $0.25 million to $1.75 million.

27. The range of responses from the individual consumers was $6 million to $7.5 million.

28. This relative lack of technical expertise occurs in such areas as equipment, available scientific man-hours, and experimental efficiency.

29. A second purpose for these telephone interviews was to pre-test questions for the Phase II survey to producers.

30. In one respect this approximation is conservative because it assumed that the quality of the industry-developed standards would have been as high as the NIST-supported ones. This was not expected to be the case, as reported in the survey.

31. This latter respondent noted that "we are in a period of transition into those areas where the technology has matured substantially. We are tending toward quality issues and will need new standards on reliability, etc. We are on another little cusp in history."

32. This assumption about a uniform decline in benefits is not critical to the analysis that follows. In present-value terms referenced to 1981, these future benefits are near zero owing to the size of the discount rate.

33. Cost data were provided by the Electronics and Electrical Engineering Laboratory at NIST.

34. The value of r that satisfies this condition is also called an internal rate of return.

35. Very Large Scale Integration.

36. See Schafft, Staton, Mandel, and Shott (1987).

37. See Schafft (undated).

38. See Schafft (1985).

39. The participating laboratories in the first (in 1986) interlaboratory electromigration experiment were Burroughs, General Electric, Hewlett-Packard, IBM, Intel, Motorola, Signetics, Syracuse University, Texas Instruments, United Technologies Microelectronics Center, and the University of Florida.

40. See Schafft, Staton, Mandel, and Shott (1987).

41. See Schafft (1987).

42. See Schafft and Albers (1988).

43. See Schafft, Lechner, Sabi, Mahaney, and Smith (1988).

44. These standards provide to industry and government, for the first time, "a set of tools to make a quantifiable assessment of the reliability of metallizations for the serious failure mechanism of electromigration in semiconductor integrated circuits. [They provide] a comprehensive statistical basis for the design, conduct, and interpretation of electromigration stress tests. [They provide] guidelines for selecting the size of the sample, the required control of the stress conditions, and the number of failures required before halting the test to attain a specific level of confidence in the results of the test. [They also provide] statistical decision rules for determining, from the test data, the relative effectiveness of metallization processes, treatments, and alloys." See Schafft (1989).

45. See Suehle and Schafft (1989, 1990).

46. See Ross (1989).

47. See Scace (1990).

48. See Scace (1990), p. 8.

49. See Tassey (undated).

50. See Engineering Research Board (1987).

51. See Augustine (1987).

52. Domestic manufacturers of semiconductors are divided into integrated or captive producers -- those that make devices for their own use -- and merchant producers -- those that primarily sell their products on the world market. IBM, AT&T, General Motors, Hewlett-Packard, Digital Equipment, and Delco are the leading captive producers of semiconductor devices. Motorola, Texas Instruments, National Semiconductor, Intel, and Advanced Micro Devices are the leading merchant producers. See Tassey (1990).

53. See U.S. Department of Commerce (April 1990), p. 93.

54. See Ross (1989).

55. This section draws from Scace (undated).

56. The storage capacity of memory devices is measured in terms of bytes (b). A byte is generally a group of eight adjacent bits and can represent one alphameric character or two digits. A bit is an acronym for binary digit, which

can be either 0 or 1.

57. See Tassey (undated).

58. See Scace (undated).

59. See Quick, Finan & Associates (1990).

60. See Quick, Finan & Associates (1990).

61. These individuals were: James Harrison, Jr. at Clemson University, Vance Tyree at the Information Science Institute of the University of Southern California, James Cottle, Jr., at the University of South Florida, Rolf Hummel at the University of Florida, and Richard Sorbello at the University of Wisconsin-Milwaukee. This is not a random sample of all academic researchers in the area of electromigration. These individuals were selected because of their knowledge of NIST's work and because their research is at the forefront of their field.

62. When a company responded with a range of cost savings, the mid-point was used for this and following calculations.

63. These are summarized in Tassey (1992).

# 6
# The Impact of External Research Relationships

Thus far, we have emphasized the effect of Federally-funded research on various aspects of firm behavior. Specifically, in Chapter 3 we developed a theoretical model to illustrate the mechanisms through which Federally-funded R&D interacts with privately-funded R&D within the research environment of the firm. Germane to that model was the role of infratechnology research. Then, in Chapters 4 and 5, we presented empirical information to illustrate the significance of infratechnology research on aspects of growth.

In this chapter we present the results of a previous empirical investigation related to external research relationships.[1] This represents only a slight change of focus from the subject matter of the previous chapters because, like those chapters, external research relationships tend to center on the expansion of the science base and other forms of nonappropriable R&D. Despite the large benefits associated with such activity, private firms will tend to underinvest in these forms of R&D because of the difficulty of appropriating fully the results. Hence, government has a valuable role in stimulating expansion of the science base.

The external research relationships emphasized are those of university-based research programs and state-funded science and technology research centers. These are especially appropriate given our overall emphasis on the role of government in innovation. University-based research is primarily basic in nature, and it is financed to a significant degree from Federal funds. Likewise, state science and technology centers, as the name implies, rely on state funding to fulfill their mission of assistance to the private center (with an end result related to regional economic development, in most cases).

After describing the data set that was examined, the results of several empirical analyses are presented. We identify empirically factors associated with the propensity of firms to engage in external research relationships and the growth consequences of such relationships.

## DESCRIPTION OF THE DATA

The data examined in this chapter come from extensive interviews with directors from both university and state research centers. From these interviews, several broad industry groups were identified to be the major "users" of external research relationships: computing equipment, machine tools, and aircraft and components. From these broad industry groups, a population of 1046 firms with 20 or more reported employees was identified from the 1986 DUNS file of the Dun and Bradstreet Corporation. After an initial mail survey to vice presidents of production/engineering, and follow-up telephone surveys, complete information was obtained on 209 firms.[2]

When surveyed, these firms were asked to classify themselves into one industry category based on their primary line of business. From their classification, these firms could be placed into five major SIC industry groups within the manufacturing sector: metalworking machinery (SIC 354), office and computing machinery (SIC 357), electronic components and accessories (SIC 367), aircraft and parts (SIC 372), and engineering and scientific instruments (SIC 381). The distribution of firms across these five industry groups is shown in Table 6.1.

The survey responses, representing 1986 activity, showed that this sample actually contains firms ranging from 7 employees (rather than 20 employees) to 390,000, with 1986 annual sales ranging from $572,000 to $34.3 billion. Table 6.2 presents the distribution of these

**Table 6.1. Distribution of Sample Firms by Industry**

| *Industry* | *n* |
|---|---|
| SIC 354 | 15 |
| SIC 357 | 69 |
| SIC 367 | 82 |
| SIC 372 | 19 |
| SIC 381 | 24 |
| | 209 |

sample firms by size category. Along with the number of firms in each size category, the average number of employees per firm is reported.

Although there are many ways to characterize the innovativeness of a firm, one dimension relates to self-financed R&D activity. The sample firms were classified as R&D-active or not based on two separate criteria: R&D expenditures and R&D personnel. *A priori*, there was no reason to believe that these two indices would be perfectly correlated. For example, a firm that relies heavily on contracted research may not have an R&D budget proportional to its R&D staff. Likewise, especially in smaller firms, the R&D budget may be so small that the category of "R&D personnel" is not meaningful, or the accounting system may not be refined sufficiently to separate R&D expenditures from other investments even when personnel are classified as being related to R&D. Nevertheless, 93 percent of the firms in this sample allocated funds to R&D in 1986 and 88 percent of all firms had at least one individual classified under the heading of R&D personnel. Table 6.3 shows the percentage of sample firms involved in R&D using each criterion. With the exception of the size category > 10,000 employees, there is a marked similarity between the percentage of firms with an R&D budget and those with classified R&D personnel. In the largest category, 93 percent of the firms reported an R&D budget but only 68 percent reported R&D

**Table 6.2. Distribution of Sample Firms by Total Employment**

| *Total Employment* | *n* | *Average Employment* | *Average Sales ($million)* |
|---|---|---|---|
| <100 | 40 | 31 | 4.33 |
| 100-249 | 83 | 118 | 11.98 |
| 250-499 | 19 | 328 | 26.81 |
| 500-999 | 17 | 653 | 57.24 |
| 1000-9999 | 22 | 2,930 | 406.37 |
| >10,000 | 28 | 76,556 | 6,259.70 |
| | 209 | | |

personnel. Perhaps, and the data do not permit an investigation of this point, the largest firms rely most heavily on contracted research. For the entire sample of firms, the correlation coefficient between total R&D expenditures and total R&D personnel is 0.653 (significant at the .01 level). Also, the percentage of the sample firms active in R&D is fairly constant across size categories, with the exception of the smallest size group, <100 employees, which had a slightly lower percentage.

Table 6.4 presents the percentage of sales devoted to R&D activity by size category for all firms in the sample. It appears that smaller firms devote a greater percentage of their sales to R&D than do larger firms. While the differences in these percentages do not appear to be significant between the middle categories, they are distinct between the category <100 employees and the category >10,000 employees. For all firms in the sample, the average percentage of sales allocated toward R&D is 10.6.

A similar pattern across size categories for all firms in the sample is shown in Table 6.5. There, the percent of total personnel

**Table 6.3. Sample Firms Reporting R&D Expenditures or R&D Employment, by Total Employment**

| *Total Employment* | *R&D Expenditures* | | *R&D Employment* | |
|---|---|---|---|---|
| | *%* | *n* | *%* | *n* |
| < 100 | 88% | 35 | 83% | 33 |
| 100-249 | 93% | 77 | 90% | 75 |
| 250-499 | 95% | 18 | 95% | 18 |
| 500-999 | 100% | 17 | 100% | 17 |
| 1000-9999 | 100% | 22 | 95% | 21 |
| > 10,000 | 93% | 26 | 68% | 19 |
| | | 195 | | 183 |

involved in R&D decreases from 16.1 percent in the category < 100 employees to 7.9 percent in the category > 10,000 employees. The variation between the middle categories is again not striking.

To summarize, five three-digit SIC industries are represented by the 209 firms in this sample. These industries are relatively research active, compared to other industries in the manufacturing sector. The sample of firms is also representative of this fact as evidenced by the relatively high percentage that reported R&D activity.

# EXTERNAL RESEARCH RELATIONSHIPS

## University-Based Research Programs

Sixty-nine percent of the sample firms were involved with a

**Table 6.4. Sample-Firm R&D Expenditures, by Total Employment**

| *Total Employment* | *R&D/Sales* | *Average R&D ($million)* |
|---|---|---|
| <100 | 13.3% | 0.49 |
| 100-249 | 10.4% | 1.14 |
| 250-499 | 12.2% | 5.82 |
| 500-999 | 12.3% | 4.23 |
| 1000-9999 | 10.5% | 35.20 |
| >10,000 | 5.0% | 417.74 |

university-based research program in 1986. As shown in Table 6.6, the degree of university involvement appears to increase with firm size. Whereas just over 50 percent of the smallest firms (less than 250 employees) were involved in a research relationship with a university in 1986, about 90 percent of the firms with more than 500 employees were active in this form of cooperative research.

Three specific categories of involvement with a university-based research program were investigated: faculty used as technical consultants (Consultants), contracted research projects (Contracts), and graduate students used as research assistants (Research Assistant).[3] The percentage of firms active in each of these three types of activities is shown in Table 6.7 by size category. In general, firms in the larger size categories make greater use of all three types of relationships than do firms in the smaller size categories. This pattern is clearest with respect to the use of university faculty as technical consultants. However, in all three cases the percentage of firms in the size category >10,000 employees who are involved in any university relationship is greater than for any of the other size categories.

Three research areas common to the underlying/emerging technology in each of the industries surveyed (based on pre-survey

**Table 6.5. Sample-Firm R&D Employment, by Total Employment**

| *Total Employment* | *Total Employment in R&D* | *Average R&D Employment* |
|---|---|---|
| <100 | 16.1% | 6.9 |
| 100-249 | 12.1% | 17.4 |
| 250-499 | 15.1% | 51.5 |
| 500-999 | 11.4% | 74.1 |
| 1000-9999 | 11.9% | 377.9 |
| >10,000 | 7.9% | 6,057.7 |

information) are artificial intelligence, computer-aided manufacturing (CAM), and computer-aided design (CAD). Each firm was asked to indicate if any of these three technology areas was a focus of their research relationship with a university. The percentages responding affirmatively, by size category, are presented in Table 6.8. Firms in the largest size category are most heavily involved in cooperative research in the field of artificial intelligence, perhaps reflecting their ability both to fund such long-term and costly research and to utilize widely whatever results from the research (economies of scope).[4] However, the same pattern is not as dramatic with regard to CAM or CAD. While firms in the category <100 employees are negligibly active in these research areas, the percentage of active firms in all of the other size categories is somewhat similar. In fact, when the subgroup of firms that produce aircraft (SIC 3721) is deleted from the analysis, the most active group of firms in cooperative activity in CAM and CAD is the group containing firms with 250 to 499 employees.

While there has been little formal empirical research on industry-university research relationships, the anecdotal evidence suggests that these type of relationships are fostered by firms for two major reasons: it is a mechanism to reduce research costs and a method to identify

**Table 6.6. Sample-Firm Involvement with University-Based Research Programs, by Total Employment**

| *Total Employment* | *%* | *n* |
|---|---|---|
| <100 | 58% | 23 |
| 100-249 | 51% | 42 |
| 250-499 | 74% | 14 |
| 500-999 | 94% | 16 |
| 1000-9999 | 86% | 19 |
| >10,000 | 100% | 28 |
| | | 142 |

productive potential employees. To investigate this issue further, each firm was asked to indicate which of the following are incentives (expected results from the relationship) for them to participate in a university-based research relationship: problem solving in production processes (PblSol), product development (PrdDev), use of university computing facilities (Compt), use of other university facilities (Facil), and gaining access to students as future employees (Emplmt). The percentage of firms noting each of these as incentives is shown in Table 6.9 by size category. With the exception of firms in the two smallest size categories, <100 employees and 100 to 249 employees, the potential for solving production process problems is instead a significant reason for firms to forge research relationships with universities. The importance of this potential as an incentive for such collaboration does not vary much by size category beyond firms with 250 or more employees. Over 60 percent of the firms in the sample, in all size categories, view product development as an important incentive for engaging in a research relationship with a university. Intercategory differences in using a university relationship as a vehicle to gain access

**Table 6.7. Sample-Firm Involvement with University-Based Research Programs, by Activity Type and Total Employment**

| *Total Employment* | *Activity Type* | | | | | |
|---|---|---|---|---|---|---|
| | *Consultant* | | *Contract* | | *Research Assistant* | |
| | *%* | *n* | *%* | *n* | *%* | *n* |
| <100 | 43% | 10 | 17% | 4 | 48% | 11 |
| 100-249 | 43% | 18 | 21% | 9 | 38% | 16 |
| 250-499 | 64% | 9 | 43% | 6 | 71% | 10 |
| 500-999 | 75% | 12 | 31% | 5 | 56% | 9 |
| 1000-9999 | 79% | 15 | 53% | 10 | 63% | 12 |
| >10,000 | 96% | 27 | 96% | 27 | 82% | 23 |

to computing facilities is primarily a small firm (<500 employees) phenomenon. It may be the case that larger firms have the in-house computer capabilities to conduct the requisite research operations. Access to other university facilities as an incentive for engaging in a university-based research relationship is important to some firms, but it does not seem to be related to the size of these firms. In accordance with anecdotal information, gaining access to students as future employees is a significant incentive for firms of all sizes to pursue university-based research relationships.

The last column in Table 6.9 reports firms' responses to a question regarding the importance of Federal tax incentives as a motivation for engaging in collaborative research (Tax) with a university. While responses vary over size categories, only in the largest size category, <10,000 employees, did more than 50 percent of the firms respond affirmatively.

**Table 6.8. Sample-Firm Involvement with University-Based Research Programs, by Research Focus and Total Employment**

| *Total Employment* | *Research Focus* | | | | | |
|---|---|---|---|---|---|---|
| | *Artificial Intelligence* | | *CAM* | | *CAD* | |
| | *%* | *n* | *%* | *n* | *%* | *n* |
| <100 | 9% | 2 | 9% | 2 | 9% | 2 |
| 100-249 | 14% | 6 | 14% | 6 | 52% | 22 |
| 250-499 | 29% | 4 | 43% | 6 | 57% | 8 |
| 500-999 | 13% | 2 | 13% | 2 | 50% | 8 |
| 1000-9999 | 26% | 5 | 26% | 5 | 37% | 7 |
| >10,000 | 71% | 20 | 57% | 16 | 61% | 17 |

Three response categories were used to determine firms' overall success with their university research relationships. The lion's share of the firms were (very or somewhat) satisfied with their collaborative research experience, as reported in Table 6.10.[5]

## State-Funded Science and Technology Center Research Programs

Sixteen percent of the sample firms were involved with a state-funded science and technology center in 1986. State centers involve the appropriation of state monies for their initiation and most operate, at least in part, on state monies. Many state-funded centers are located on university campuses, but some are adjacent to a university campus with

**Table 6.9. Sample-Firm Incentives to Engage in University Research Relationships, by Total Employment**

| *Total Employment* | *Incentives* | | | | | | | | | | | |
|---|---|---|---|---|---|---|---|---|---|---|---|---|
| | *PblSol* | | *PrdDev* | | *Compt* | | *Facil* | | *Emplmt* | | *Tax* | |
| | *%* | *n* | *%* | *n* | *%* | *n* | *%* | *n* | *%* | *n* | *%* | *n* |
| <100 | 4% | 1 | 61% | 14 | 4% | 1 | 22% | 5 | 57% | 13 | 22% | 5 |
| 100-249 | 19% | 8 | 69% | 29 | 19% | 8 | 19% | 8 | 64% | 27 | 17% | 7 |
| 250-499 | 57% | 8 | 71% | 10 | 29% | 4 | 57% | 8 | 79% | 11 | 43% | 6 |
| 500-999 | 38% | 6 | 63% | 10 | 6% | 1 | 19% | 3 | 69% | 11 | 31% | 5 |
| 1000-9999 | 42% | 8 | 47% | 9 | 5% | 1 | 16% | 3 | 84% | 16 | 11% | 2 |
| >10,000 | 61% | 17 | 79% | 22 | 14% | 4 | 61% | 17 | 93% | 26 | 54% | 15 |

**Table 6.10. Sample-Firm Satisfaction in University Research Relationships, by Total Employment**

| *Total Employment* | *Level of Satisfaction* | | | | | |
|---|---|---|---|---|---|---|
| | *"Very Satisfied"* | | *"Somewhat Satisfied"* | | *"Not Satisfied"* | |
| | *%* | *n* | *%* | *n* | *%* | *n* |
| <100 | 30% | 7 | 65% | 15 | 5% | 1 |
| 100-249 | 38% | 16 | 62% | 26 | 0% | 0 |
| 250-499 | 71% | 10 | 29% | 4 | 0% | 0 |
| 500-999 | 44% | 7 | 56% | 9 | 0% | 0 |
| 1000-9999 | 47% | 9 | 53% | 10 | 0% | 0 |
| >10,000 | 25% | 7 | 75% | 21 | 0% | 0 |

independent political authority. The general pattern seems to be that states appropriate monies for the center's infrastructure and some of its operating costs, while individual research contracts are negotiated between individual centers or universities and industrial companies on a proprietary basis.

A comparison of the data in Table 6.11 with that in Table 6.6 shows that firms in all size categories are more likely to be involved in a cooperative research program with a university than with a state-funded science and technology center. As with university-based research, the likelihood of collaboration increases with firm size.

These proprietary relationships with state centers cover a wide range. The first proprietary-relationship category in Table 6.12 reports the percentage of firms responding that their involvement with a state center led to the development of prototype products (PrdDev). On average, this result occurred more frequently in small firms (<500

**Table 6.11. Sample-Firm Involvement with State-Funded Science and Technology Centers, by Total Employment**

| *Total Employment* | *%* | *n* |
|---|---|---|
| < 100 | 5% | 2 |
| 100-249 | 8% | 7 |
| 250-499 | 16% | 3 |
| 500-999 | 18% | 3 |
| 1000-9999 | 18% | 4 |
| > 10,000 | 54% | 15 |
| | | 34 |

employees) than in large firms (> 499 employees). For relationships leading to improvements in production processes (PrdPrc), no cross-category differences could be found. Each firm was also asked if their participation in center activities stimulated research within their company (R&D). The summary data reported in the R&D column indicate that this was the case in most instances. The exception was firms in the size category < 100 employees. The data reported for the fourth proprietary-relationship category summarize firms' responses to a question related to whether their participation in a center program had any effect on improvements in the quality of products and services (Qlty). Small firms responded in the affirmative to this inquiry more than twice as often as did large firms. Some state-center collaborative research did lead to a reduction in the price of products to the firms' customers (Price), as shown for the fifth category, but there is great variation across firms regarding the incidence of this. State science center relationships led more frequently to a reduction in production costs (Cost) in firms in the smaller size categories than in firms in the larger size categories. The data for the seventh proprietary-relationship category indicate clearly that

**Table 6.12. Results of Research Relationships with State-Funded Science and Technology Centers for Sample Firms, by Total Employment**

| *Total Employment* | *Results* | | | | | | | | | | | | | | | |
|---|---|---|---|---|---|---|---|---|---|---|---|---|---|---|---|---|
| | *PrdDev* | | *PrdPrc* | | *R&D* | | *Qlty* | | *Price* | | *Cost* | | *Div* | | *Eval* | |
| | *%* | *n* | *%* | *n* | *%* | *n* | *%* | *n* | *%* | *n* | *%* | *n* | *%* | *n* | *%* | *n* |
| <100 | 50% | 1 | 50% | 1 | 50% | 1 | 100% | 2 | 100% | 2 | 50% | 1 | 50% | 1 | 50% | 1 |
| 100-249 | 43% | 3 | 43% | 3 | 86% | 6 | 100% | 7 | 14% | 1 | 43% | 3 | 43% | 3 | 43% | 3 |
| 250-499 | 100% | 3 | 33% | 1 | 100% | 3 | 33% | 1 | 33% | 1 | 67% | 2 | 67% | 2 | 0% | 0 |
| 500-999 | 33% | 1 | 33% | 1 | 67% | 2 | 33% | 1 | 33% | 1 | 33% | 1 | 0% | 0 | 0% | 0 |
| 1000-9999 | 25% | 1 | 25% | 1 | 100% | 4 | 50% | 2 | 0% | 0 | 25% | 1 | 0% | 0 | 0% | 0 |
| >10,000 | 33% | 5 | 40% | 6 | 80% | 12 | 27% | 4 | 33% | 5 | 40% | 6 | 13% | 2 | 13% | 2 |

a primary result of state center relationships for small firms is an increased diversity of products (Div). The information presented for the last category shows that firms in the smaller size categories obtain information from these relationships that helps them develop criteria and methods for project evaluation (Eval). There is little evidence that firms in the larger size category view this as a significant outcome from state center collaboration.

A comparison of the data in Table 6.13 with that in Table 6.10 shows that a greater percentage of participants in state-funded science and technology center research relationships were not satisfied, compared to their involvement with universities.[6] This dissatisfaction occurred primarily among small firms.

Eleven separate categories of the use of external research relationships (university and state science centers) were identified in the survey. Three of these categories were discussed in Table 6.7 with regard to university relationships and eight were discussed in Table 6.12 with regard to state center relationships. A utilization of technical expertise index was created from these primary data. This firm-specific index represents the percentage of all of the areas of possible technical expertise from an external research relationship utilized by a firm. This variable (Utilization) has a minimum value of one-eleventh -- each of the 142 (68 percent of 209) externally-research involved firms reported at least one area of utilization -- and a maximum value of unity. The mean values of this utilization rate variable are reported in Table 6.14 by size category. An inspection of these data suggests that there is little cross-size category variation in the utilization of technical expertise by firms until the largest category, >10,000 employees.

This overview of the primary data suggests several preliminary patterns of firm behavior.

One, firms in all size categories are more likely to participate in a cooperative research relationship with a university than with a state science and technology center. However, firms in the larger size categories are more likely to participate in both (Table 6.6 and Table 6.11).

Two, firms in all size categories were engaged in university research relationships to use faculty as technical consultants. Firms in the larger size categories do this to a greater degree than firms in the smaller size categories (Table 6.7). This collaboration tends to be oriented primarily toward product development and secondarily toward problem solving in areas related to production (Table 6.9).

**Table 6.13. Sample-Firm Success in State-Funded Science and Technology Center Research Relationships, by Total Employment**

| *Total Employment* | *Level of Satisfaction* | | | | | |
|---|---|---|---|---|---|---|
| | *"Very Satisfied"* | | *"Somewhat Satisfied"* | | *"Not Satisfied"* | |
| | *%* | *n* | *%* | *n* | *%* | *n* |
| <100 | 0% | 0 | 50% | 1 | 50% | 1 |
| 100-249 | 29% | 2 | 57% | 4 | 14% | 1 |
| 250-499 | 33% | 1 | 67% | 2 | 0% | 0 |
| 500-999 | 33% | 1 | 67% | 2 | 0% | 0 |
| 1000-9999 | 50% | 2 | 50% | 2 | 0% | 0 |
| >10,000 | 67% | 10 | 20% | 3 | 13% | 2 |

Three, except for firms with <100 employees, a common area of collaboration with universities was computer-aided design (Table 6.8).

Four, in addition to research expertise, firms in all size categories viewed access to students as future employees as a significant incentive for engaging in a university-based research relationship (Table 6.9).

Five, firms in small size categories reported that their cooperative research relationships with state science and technology centers have, on average, led to internal product (quality and diversity) and process (Cost) improvements more often than did firms in larger size categories (Table 6.12).

Six, firms in all size categories appeared to be pleased with their research relationships with both university and state centers (Table 6.10 and Table 6.13).

Seven, except for firms in the very largest size category

**Table 6.14. Sample-Firm Utilization of Technical Expertise from External Relationships, by Total Employment**

| *Total Employment* | *Average Utilization* |
|---|---|
| < 100 | 0.27 |
| 100-249 | 0.28 |
| 250-499 | 0.27 |
| 500-999 | 0.20 |
| 1000-9999 | 0.22 |
| > 10,000 | 0.42 |

(> 10,000 employees), size does not appear to be an important factor in explaining interfirm differences to the extent to which the technical expertise available from external research relationships is utilized by a firm (Table 6.14).

## EMPIRICAL ANALYSIS

### Explaining Utilization of External Research

The data presented in Table 6.14 show that the average utilization rate (percentage of identified areas through which a firm utilized technical expertise from an external research relationship) is larger among firms in the size category > 10,000 employees than in any other size category. This may suggest that, beyond a threshold level, size is an important correlate with a firm's ability to internalize the economic benefits associated with external research relationships.

To test this hypothesis, the following regression model was estimated:

$$\text{UTILIZ} = \alpha + \beta_0\ \text{SIZE} + \beta_1\ \text{BASRES} + \beta_2\ \text{Ds} + \epsilon$$

where UTILIZ is the firm-specific utilization rate, SIZE is measured as the number of employees (1000s), BASRES is a binary variable equalling 1 if the firm was active in basic research (all firms are active in R&D) and 0 otherwise, D is a vector of three-digit SIC industry dummy variables, and $\epsilon$ is a normally distributed random error term.

*A priori*, it is hypothesized that the estimated coefficient on BASRES is positive. To the extent that activity at the basic research end of the R&D spectrum proxies the diversity of a firm's technical activities/needs, basic research-active firms may have the greater technical capacity to internalize the results from research cooperation with universities and/or science centers.[7]

The least-squares regression results are presented in Table 6.15. The estimated coefficient on SIZE in column (1) is not statistically significant. Firms involved in basic research utilize a significantly greater percentage of potential technical benefits from external research relationships than do firms who are not. Also, to a large degree, interfirm differences in the utilization of technical expertise is industry specific as evidenced by the individual significance of the industry dummies. As a group, these coefficients are significantly different from zero, too. Also, the results presented in column (2) do not suggest any non-linear size effect, as could have been interpreted from the descriptive data in Table 6.14. Industry concentration ratios and industry indices of international competition were included in separate regressions, but their estimated coefficients were not significant at a conventional level.

## External Research Relationships and the Size Elasticity of R&D

One of the most frequently investigated topics in the economics of technological change literature is the relationship between firm size and the corresponding level of R&D activity. The question often asked is, "Do large firms spend more on R&D relative to their size than do small firms?" This question is reinvestigated here using the sample data described above. R&D activity is measured in terms of R&D expenditures and firm size is measured as total sales.

Using a double-log regression specification:[8]

**Table 6.15. Least-Squares Regression Results Explaining Interfirm Differences in the Utilization of Technical Expertise from External Research Relationships**

| *Variables* | *Estimated Coefficients* | |
|---|---|---|
| | *(1)* | *(2)* |
| Constant | 0.325<br>(12.42) | 0.319<br>(12.16) |
| SIZE | 0.000009<br>(0.03) | 0.01<br>(1.56) |
| SIZESQ | --- | -0.000004<br>(-1.46) |
| BASRES | 0.079<br>(2.57) | 0.056<br>(2.01) |
| D354 | -0.093<br>(-1.76) | -0.086<br>(-1.62) |
| D357 | -0.112<br>(-3.36) | -0.106<br>(-3.16) |
| D367 | -0.054<br>(-1.69) | -0.046<br>(-1.43) |
| D372 | -0.102<br>(-2.03) | -0.102<br>(-2.04) |
| $R^2$ | 0.1234 | 0.1357 |

*Note:* t-statistics reported in parentheses; n = 142.

$$\ln R\&D = \alpha + \beta \ln SIZE + \epsilon$$

the estimated value of $\beta$ is the size elasticity of R&D. The estimated regression results, with t-statistics reported in parentheses, are:

$$\ln R\&D = \underset{(6.77)}{2.46} + \underset{(26.97)}{0.93} \ln SIZE$$

$$R^2 = 0.797.$$

The estimated elasticity of 0.93 is significantly different from zero, but is not significantly different from unity.

To investigate for differences in the size elasticity of R&D between small and large firms, and for the extent to which these differences (if any) are influenced by external research activity, the sample of firms was divided into a subsample of small firms (<500 employees) and a subsample of large firms (>499 employees). For each subsample, the following regression model was estimated:

$$\ln R\&D = \alpha + \beta_0 \ln SIZE + \beta_1 D \ln SIZE + \epsilon$$

where D equals 1 for firms involved with university-based research programs and 0 otherwise. The size elasticity of R&D for firms not involved with a university is $\beta_0$, and the elasticity for firms involved with a university is $(\beta_0 + \beta_1)$. The calculated elasticities from estimating separately the above equation for the small firm subsample and for the large firm subsample are reported in Table 6.16.

Small firms are more responsive to a percentage change in sales in terms of their R&D spending than are large firms.[9] There is no significant difference between the R&D elasticity in large firms who are and who are not involved in cooperative research with a university (0.87 versus 0.88). However, in the subsample of small firms, the estimated elasticity is significantly greater for those firms involved with university-based research than for those not so involved (0.99 versus 0.93).[10]

We interpret the results in Table 6.16 to mean that small firms are able to use their external research relationships with universities to gain flexibility in their internal R&D activity. To the extent that this interpretation is valid, it may explain why small firms are more innovative, relative to their size, than large firms. To the extent that there are innovation-based diseconomies of scale in large firms (owing to bureaucratization in the innovation decisionmaking process), then small firms may avoid this through reliance on university-based sources of research-related knowledge.

Regarding state-funded science and technology centers, there was no empirical evidence that external research relationships affected firms'

**Table 6.16. Estimated Size Elasticities of R&D**

| *Involvement with Universities* | *Small Firms* | *Large Firms* |
|---|---|---|
| Involved | 0.99 | 0.87 |
| Not Involved | 0.93 | 0.88 |

*Note:* Calculated elasticities are significantly different from zero but not significantly different from unity.

size elasticity of R&D spending. As noted in Table 6.12 above, firms involved with state centers do realize technical benefits, but apparently these do not affect their R&D responsiveness.

## External Research Relationships and the Efficiency of R&D

As already explained, a framework frequently used by researchers for estimating the returns to R&D spending reduces to the following regression model:

$$TFP = \alpha + \beta\,(RD/S) + \epsilon$$

where TFP represents total factor productivity, (RD/S) is the ratio of R&D spending to firm sales, and $\beta$ is the estimated rate of return to R&D.

To estimate this model, data were needed for the calculation of total factor productivity over a defined time period. Sufficient data for these calculations were not available for all firms in the sample. The critical data element for the calculation of total factor productivity is an estimate of each firm's capital stock. This information was unavailable for a number of firms, although in some instances secondary data (e.g., Compustat) were used. TFP over the period 1982 to 1987 could be calculated for only 158 of the 209 firms in the sample. The 51 firms deleted from this analysis were mostly small firms with <100 employees.[11]

Several versions of the basic rate of return model were estimated. First, in order to test for differences between the rate of return to R&D in large versus small firms, a second regressor was included in the above equation. This regressor equalled $\beta_2 D(RD/S)$ for D a binary variable equalling 1 for small firms (<500 employees) and 0 otherwise. The estimated least-squares coefficient on this term was not statistically different from zero, implying that there was no statistical difference between the rate of return to R&D in firms in the two size groups.[12] Overall, the rate of return to the 158 firms in the subsample was 26.1 percent.[13] This result is shown in Table 6.17.

Second, a similar specification was estimated to account for possible differences in the rate of return to firms engaged in and not engaged in external research relationships (with a university or a science center, or both). For this, the variable D above equalled 1 if the firm was so engaged and 0 otherwise. As reported in Table 6.17, the estimated returns to R&D in firms involved in external research relationships are more than twice those of firms that are not -- 34.5 percent versus 13.2 percent.[14]

Finally, segmenting both by size and by involvement in an external research relationship by including two regressors in the original specification (one with a size dummy and the other with an external research relationship dummy), the return to R&D in small, externally-research-active firms was found to be quite large. As reported in Table 6.17, the estimated rate of return to R&D in this group of firms was 44.0 percent compared to 29.7 percent in large externally-research-active firms. The equivalent rates for large and small firms not involved in external research was also much lower (approximately 14 percent). Therefore, small firms appear to be able to transfer knowledge gained from external research relationships to increase the efficiency of their R&D activities and more effectively than large firms.

## CONCLUSIONS

As we stated in Chapter 1, a role of government is to fund basic research, especially at universities, and to encourage industry-university relationships. The descriptive data presented in this chapter clearly

**Table 6.17. Estimated Rates of Return to R&D Expenditures**

| *Involvement in External Research* | *Small Firms* | *Large Firms* | *All Firms* |
|---|---|---|---|
| All Firms | 26.1% | 26.0% | 26.1% |
| Involved | 44.0% | 29.7% | 34.5% |
| Not Involved | 13.2% | 14.1% | 13.9% |

indicate that there are benefits to companies that interact with university-based research centers (and state-based science and technology centers). Although we are unable to evaluate whether an optimal level of Federal support is going to universities, we are able to conclude that the current level of commitment seems to have productivity-related benefits to those that access these centers.

## NOTES

1. This chapter is based on research funded by the U.S. Small Business Administration, Office of Advocacy. An earlier version of parts of this chapter appeared in Link and Rees (1990).

2. The 1046 firms were divided between these industries as follows: 295 from computing equipment, 207 from machine tools, 390 from electronic components, and 154 from aircraft and components. Whenever possible, reported survey information (e.g., sales data) was verified against published data (e.g., Form 10-K data) to insure response reliability. When explainable differences occurred (e.g., a survey respondent reporting sales in $millions rather than $thousands) the primary data were corrected.

3. No information was collected regarding whether this involvement occurred at the firm or at the university.

4. Current concern with artificial intelligence can also be seen as an evolution from past concern with CAD/CAM systems and other aspects of flexible manufacturing. See Rees (1986).

5. There is not sufficient variation between the three response categories to conduct a more detailed investigation of interfirm differences in success with university-based research.

6. All firms participating in a state-funded science and technology research relationship were involved in a university-based research relationship. As with university-based research, there is not sufficient variation between response categories to conduct a more detailed investigation of inter-firm differences in success with state science center research.

7. Nelson (1959) is credited with the hypothesis that diversity is a necessary condition for firms to undertake the uncertainty associated with basic research. See also Link and Long (1981).

8. Link, Seaks, and Woodbery (1988) verify the appropriateness of this double-log specification.

9. Finer size categories were investigated, but little additional information was obtained from subdividing either subsample. As reported in Table 6.6, all of the sample firms with >10,000 employees were involved with a university research program.

10. These results are unchanged when SIZE is measured in terms of number of employees.

11. The 158 retained firms were divided by size category as follows: 9 firms with <100 employees, 74 firms with 100-249 employees, 16 firms with 250-499 employees, 15 firms with 500-999 employees, 20 firms with 1000-9999 employees, and 24 firms with >10,000 employees.

12. Link's (1980) analysis of the rate of return to R&D among firms in the chemicals industry found that the return increased with firms size to a modest threshold level, and then it remained constant.

13. Recall that the firms in this study were drawn from some of the more R&D-active industries in manufacturing.

14. This finding corresponds favorably with that of Link and Bauer (1989). They report, based on a sample of 92 manufacturing firms, the rate of return to R&D in firms involved in cooperative research programs with other firms is nearly three time that of firms not so involved -- 37.7 percent versus 12.9 percent.

# 7
# Policy Initiatives to Support Innovation

As we discussed in Chapter 1, governmental support of innovation can be either direct or indirect. The primary indirect mechanisms used by government to stimulate innovation have historically been tied to tax policies. However, the descriptive data in Chapter 6 suggested that Federal tax incentives as a motivation for engaging in collaborative research with a university was not generally viewed as an effective policy tool.

The R&D tax credit is scrutinized in this chapter. Initially heralded by the Reagan Administration, the evidence to date on the effectiveness of the tax credit is mixed. Unlike the previously discussed efforts by government to support innovation, the R&D tax credit is less focused and thus will not necessarily aid the important nonproprietary categories of activities that support innovation. As with any R&D-focused governmental policy whose effect is to stimulate proprietary (i.e., appropriable) research, the R&D tax credit is therefore less desirable than the direct mechanisms previously discussed in this book. Any policy aimed at R&D *per se* that is not complemented with policies to leverage the efficiency of R&D will be less effective than desired.

See again Figure 1.1 for a schematic representation that emphasizes this issue.

## THE RESEARCH AND DEVELOPMENT TAX CREDIT

### Tax Incentives as a Policy Tool[1]

There are two fundamental questions which need to be considered when analyzing any policy tool: the question of need and the question of tool selection. For the R&D tax credit, the questions are: Is there a need for a policy tool to stimulate R&D? and, Are tax incentives the appropriate tool for stimulating R&D?

The question of need has been debated for some time. As reviewed in the first chapter, however, the theoretical and empirical evidence seems to suggest that there is indeed a need. Given this conclusion, the next question to ask whether tax incentives are preferable to other policy tools, such as direct grants assistance. There are compelling arguments for and against the use of tax incentives to achieve the policy goal of stimulating R&D. The advantages and disadvantages of tax incentives *per se*, compared to grants assistance programs in particular, are discussed below.

**Advantages of Tax Incentives.** We see at least six advantages to tax incentives, especially those related to R&D.

(1) *Tax incentives entail less interference in the marketplace than grants assistance, and relatedly, allow more private decisionmakers to retain autonomy*. It has been claimed that "tax incentives [for R&D expenditures] do not require that governmental officials make difficult, subjective judgments concerning the relative results of various innovations and technologies," and that tax incentives do not create artificial markets because "firms are still free to design, price and sell in response to real demand, rather than in response to government-created demand."[2]

It is difficult to find fault with the first point, at least from an empirical standpoint. Tax incentives do seem to provide more discretion for private decisionmakers. However, it is not clear that there is any necessity for grants assistance programs to encumber private decisionmaking because it is theoretically possible to devise grants assistance and subsidy programs which entail very few controls.[3] By the same token, it is possible to formulate tax incentive programs which specify so many qualifying criteria that the effects are quite similar to a tightly controlled grants assistance program. Nevertheless, it is true that tax incentive programs tend to rely more on the discretion of private decisionmakers and less on the preferences of government officials.

(2) *Tax incentives require less paperwork than grants assistance programs and thus fewer layers of bureaucracy.* The red tape, paperwork, and additional accounting associated with many grants assistance programs are all too familiar. Tax incentives do require administration and paperwork, but the administrative burden is significantly less. Furthermore, governmental bureaucracy (e.g., the Internal Revenue Service in the United States) is already established and well trained in auditing and tax administration. Also, there is an administrative advantage to tax incentives over grants assistance programs because the rules of the game change less rapidly. One of the greatest boons to this administrative practice is predictability. Policy following from tax incentives is more predictable and stable than policies which require yearly appropriations and are subject to rapid legislative changes.

Some take issue with this claim that tax incentives require less government interference contending that there is little value in comparing a direct assistance policy that is tightly controlled to a tax incentive policy that has almost no controls.[4] If controls are the cutting issue, it is possible to devise grants assistance programs that entail almost no controls, just as it is possible to devise tightly controlled tax incentive policies. The same argument can be advanced in connection with the alleged advantages of tax incentives in promoting private decisionmaking. Moreover, the fact that the existing tax incentives are indeed less structured than most direct expenditure programs is a virtue limited to the eye of the beholder. In some instances, lack of structure in tax incentives represents little more than a retreat from planning and decisionmaking responsibility.

(3) *Tax incentives, compared to grants assistance programs, avoid the need to make difficult distinctions or to set nebulous and particularistic requirements for receiving assistance.* There is an administrative benefit to being able to avoid particularistic requirements

with regard to efficiency and equity. With regard to R&D, tax incentives are ideally designed to reward past behavior. This implies that past behavior has been successful and is worth rewarding.

(4) *Tax incentives have the psychological advantage of achieving a favorable industry reaction vis-a-vis grants assistance programs.* The supposed psychological advantages of tax incentives vis-a-vis grants assistance programs are difficult to document. Some argue that, irrational as it may be, people do react differently to tax incentives than to direct assistance. Tax incentives seem to draw nourishment from the taproot of the free enterprise symbol.[5] The free enterprise ethic has considerable symbolic capacity, and it would be surprising not to find a differential reaction among businessmen. Of course, the real question is the behavioral significance of these (possibly) different reactions. If businessmen were more likely to eschew, for psychological reasons, money provided via assistance programs, this difference would be important.

(5) *Tax incentives are more permanent and stable than grants assistance programs in that they do not require an annual budget review.* Firms are more likely to make fundamental changes in their plans and investment strategies if they perceive that a policy has some stability. Thus, a tax credit which has a known long-term or open-ended effective life will be more likely to lead to adjustments in behavior than a grants assistance program that is here today but may be gone with tomorrow's round of budget cutting. This seems to be one of the most powerful arguments for permanent tax incentives as a policy tool for shaping R&D activity.

(6) *Tax incentives, compared to grants assistance programs, have a high degree of political feasibility.* It is often the case that tax incentives face less political opposition than direct grants assistance policies. What better evidence of this is there than the rapid growth of the tax expenditures budget, even as officials scour budgetary lines for possible spending reductions. Also, there is a different constituency for tax incentives than for grants assistance programs. Tax incentives are typically the favored child of conservative politicians and their business constituencies, not only in the United States but also in other industrialized nations. One of the first steps that newly-elected conservative administrations take is to expand tax incentives: one of the

first steps that liberal administrations take (or at least give lip service to) is to close loopholes. A very pragmatic argument for tax incentives for R&D expenditures is (to borrow a worn cliche) that one should not look a gift horse in the mouth. If conservative Republican administrations choose to use this means of encouraging R&D spending, why resist? By the same token, when tax incentives go out of fashion we can (to mix equine metaphors) switch horses in the middle of the race and advocate some other politically feasible approach to enhancing R&D spending.

**Disadvantages of Tax Incentives.** Of course, critics of tax incentives may disagree with these alleged advantages and the implied criteria against which they are made. Perhaps even more compelling than simple disagreements are the critics' arguments for the disadvantages of a tax incentive approach. We note five.

(1) *Tax incentives, unlike grants assistance programs, bring about unintended windfalls by rewarding what would have been done without the tax incentive.* This proposition requires, for strict verification, knowledge which in principle is unobtainable. Once a tax incentive is actually introduced, we can never know with certainty what behavior would have been had it not been introduced. But, the logical dilemma has not curtailed speculation or *post hoc* analysis.

In many instances, windfalls are inefficiencies rather than intended consequences of policies. Windfalls can be avoided simply by introducing adequate controls and qualifications. However, so doing introduces complexities and, moreover, resurrects the very problem that tax incentives are supposed to overcome: government intervention. Alternatively, the magnitude of the windfall can be curtailed if the tax credit is initiated on an incremental basis.

(2) *Tax incentives, unlike grants assistance programs, often result in undesirable inequities.* Tax incentives are likely not only to introduce inequities but also to mask those inequities.[6] The problem is that tax incentives are a blunt instrument and that highly specific provisions would have to be written into the tax code to deal effectively with inequities.

It is not simply that differential rewards are provided under tax incentives -- indeed differential rewards are often undesirable -- but there is a tendency for lopsided benefits as well. Typically, taxpayers in higher income brackets benefit the most from tax incentives. This is especially a problem in the case of tax incentives for R&D because many new firms have no tax liability (i.e., taxes to be paid) and are not profitable during the early years in which they develop products and initially invest in R&D assets. But, in many cases, those are the very firms which most often warrant assistance. Carry-forward provisions (i.e., using tax deductions in some future year(s) rather than in the first eligible year) can in some instances begin to address this most troublesome of inequities; however, even then, there is potential for "corporate regressivity" in tax incentives. Tax incentives as a policy tool toward R&D are most effective when they are least necessary and may influence those firms who need them the least.

(3) *Tax incentives, unlike grants assistance programs, raid the Federal treasury in the sense that tax rates end up higher than they would otherwise be.* Tax expenditures represent foregone revenue and, *ceteris paribus*, the tax rate would have to be elevated just to maintain revenues. However, all else is not equal because revenue is not simply a function of tax rates. Supply-side arguments tell us that tax expenditures can ultimately lead to an increase in revenues provided that the incentives lead ultimately to productivity growth.[7] Nevertheless, it is important to note that tax incentives are usually open-ended with no limit or maximum amount on the dollar value of the credit a taxpayer can earn. There is an unpredictable element to any tax incentives because the effects are determined by a great many economic and political variables.

(4) *Tax incentives, unlike grants assistance programs, often undermine budget controls and public accountability.* The stabilizing influence provided by tax incentives vis-a-vis grants assistance is not without costs. Congress has essentially forfeited its oversight function in providing assistance through the tax credit policy tool. This has a number of ramifications. In the first place, tax incentives are awarded a status not enjoyed by other governmental initiatives because they are off budget (i.e., not part of the annual itemized budget review process) and therefore an uncontrollable expenditure. This not only has the effect of reducing control and coordination of policy, but also of shifting the locus of controls away from the substantive committees and subcommittees designed to authorize the budget -- the bodies that should have the most expertise and responsibility for R&D policy -- to the tax committees (the Senate Finance Committee and the House Ways and Means Committee). Congress passed the Congressional Budget and

Impoundment Control Act of 1974 in part as a means to exert managerial control of the budget and to eliminate backdoor spending. Tax expenditures are an especially pernicious form of backdoor spending because the revenue committees themselves (rather than the authorizing committees) are opening the backdoor. Unless there is a clear conviction that policies implemented via tax expenditures merit an immunity not granted to other R&D assistance programs, the result is an unnecessary abrogation of policy leadership. In an era of budget cutting, any step that limits the ability of Congress to manage its policy priorities should be given close scrutiny.

(5) *The effectiveness of tax incentives, unlike grants assistance programs, in stimulating R&D varies over the product life cycle.* It is not always in the best interest of a firm to invest additional R&D in product or process technology. Over the product life cycle, internal R&D is most effective during the growth and maturity stages. During the introduction and decline stages of the life cycle, firms will tend to rely on purchased technology (i.e., general technology developed by other companies).[8] Thus, grants assistance programs become relatively more effective as a policy tool for stimulating innovation-based efficiency during these two stages.

## Tax Policies Affecting R&D

The adoption of Section 174 of the Internal Revenue Code in 1954 codified and expanded tax laws pertaining to firms' R&D expenditures. This provision permits businesses to deduct fully research and experimental (R&E) expenditures, but not development or research application expenditures, in the year incurred.[9] There is also the option to capitalize R&E expenditures and amortize them over a period of not less than five years, beginning with the month in which benefits are first realized. By assumption, these benefits occur in the month in which the product or process gained from the R&E activity first produces income. This deferral option was intended to benefit the small newer firms who had little or no taxable income during their early years.

The Internal Revenue Service made clear that the expensing option was limited to R&E expenditures and was not applicable to expenditures on capital assets necessary for the conduct of R&E activities. However, the costs of such depreciable assets as machinery, equipment and facilities is partly recovered through depreciation allowances that apply to investments in any depreciable property.[10]

Expenditures for R&E are given preferential treatment under Section 174, but there was little indication initially regarding those activities which qualified as R&E. Stipulations and limits were provided in subsequent Department of Treasury regulations. Research and development cost in the experimental or laboratory sense include all such expenditures that are incident to the development of an experimental or pilot model, a plant process, a product, a formula, an invention or similar property, and improvements to existing property similar to these types. As well, it includes the cost of obtaining a patent, such as attorney fees incurred in making or perfecting a patent application. Specifically excluded are expenditures for testing or inspection of materials or products for quality control, management studies, advertising and promotion. A number of court decisions have further codified the legal meaning of research and experimental expenditures.

While the immediate expensing of R&E expenditures is generally viewed as a tax incentive, it has been argued that Section 174 was more of an administrative convenience than a *per se* tax incentive.[11] While there is a consensus about the salutary effects of R&E expenditures on both innovation and the creation of marketable products, the marginal contribution of R&E is difficult to determine as it is closely intertwined with a host of other inputs such as management structure, quality control, marketing strategy, and level of employee creativity.

A number of additional elements, other than Section 174, of the Tax Code are generally viewed as R&D incentives.[12] Individuals and corporations are allowed to deduct, according to Section 170(a), contributions to educational and scientific organizations held to be operating in the public interest. There are limits on such deductions: individuals may deduct not more than 50 percent of adjusted gross income for such contributions and corporate deductions are limited to 5 percent of taxable income. Furthermore, the income of scientific and educational organizations operated in the public interest is exempt from Federal income tax. According to Sections 501(a-c), organizations qualify for this exemption if they are conducting scientific research that is "directed toward benefiting the public." The operating standard is that the work must result in information "published in a treatise, thesis, trade publication, or in any other form that is available to the interested public." If met, the organization qualifies for the exemption even if the research is performed under "a contract or agreement under which the sponsor or sponsors of the research have the right to obtain ownership or control of any patents, copyrights, processes, or formulae resulting from such research."

One of the most significant, and in some cases controversial, changes in the tax treatment of R&E expenditures and R&E assets was provided in the Reagan Administration policy centerpiece, the Economic Recovery Tax Act of 1981 (ERTA). The four major changes embodied in ERTA include a faster depreciation of R&E assets, a two year suspension of Treasury Regulation 1.861-8 to study its impact,[13] an increase in the deduction promoted for contributions of newly manufactured research equipment to universities, and a tax credit for R&E expenditures.

Because our primary interest is the tax credit for R&E, we focus here specifically on the last provision in ERTA. The credit is for qualified R&E expenditures in excess of the average amount spent during the previous three taxable years or 50 percent of current year expenditures, whichever is greater. Expenditures qualifying include company-financed expenditures for R&E wages and supplies, 65 percent of the amount paid for contracted research, and 65 percent of corporate grants to universities and scientific research organizations for basic research. Expenses must be paid by the taxpayer during the taxable year and must pertain to the carrying on of a trade or business. Thus, the credit was not available to start-up companies, certain joint ventures, or to existing firms entering into new lines of business. The credit became effective for R&E expenditures paid or incurred after June 30, 1981, and before January 1, 1986, when the credit would have expired.

The tax credit was 25 percent of the excess of the taxpayer's qualified research for the taxable year over the average of the taxpayer's annual research expenditures during the base period. The base period was generally defined to be the three years immediately preceding the taxable year for which the credit is claimed. For 1981, the base was 50 percent of 1980 qualified expenditures (because only spending during the second half of 1980 was eligible); for 1982 it was the mean of 1980 and 1981 qualified expenditures; and for 1983, 1984, and 1985 the base was the mean of the previous three years.[14] However, the base period expenditures could not be less than 50 percent of qualified expenditures for the current year.[15]

Several limitations are worth noting. The requirement for "carrying on a trade or business" means that expenses incurred in connection with trade or business but not pertaining to development of potentially marketable goods and services failed to qualify. Perhaps just as important, the credit did not apply to research expenditures paid or incurred prior to commencing a trade or business. Only wages paid for doing actual research work qualified for the credit. Thus, wages for laboratory scientists and engineers and their immediate supervisors qualified, but wages for general administrative personnel or other auxiliary personnel (such as computer technicians working in a multipurpose computer and information processing department) did not. Also, research done outside of the United States was excluded from the operable definition. It has been estimated that because of these limitations only about 70 percent of total company-financed R&D qualified for the credit between 1981 and 1985.[16]

The Tax Reform Act of 1986 reduced the legislated credit rate from 25 percent to 20 percent.[17] Also, the definition of qualified research was narrowed in order to emphasize the intended technological nature of the targeted research activities. It was believed that the credit had been claimed for virtually any product development expense whether innovation was involved or not. The intention of narrowing the definition of qualified expenditures was to eliminate product modifications after they reached their functional specifications. By so doing, the credit should focus on the discovery and experimentation stages of the innovation process.[18]

The Basic Research Credit was also included in the Tax Reform Act of 1986. This credit was intended to encourage corporate support of university-based and nonprofit-based basic research. It is set at a fixed rate of 20 percent. Thus, it provides a greater tax advantage than does the conduct of basic research in-house under the R&E tax credit, *ceteris paribus*.

The Technical and Miscellaneous Revenue Act of 1988 (TAMRA), Section 280C(c) in particular, disallowed deductions for the portion of qualified R&D that equalled 50 percent of the taxpayer's research credit. According to the U.S. Department of Commerce, Office of Technology Policy, this change reduced the maximum effective rate from 20 percent to 16.6 percent.[19] Also, this act mandated that the General Accounting Office (GAO) must submit to Congress a report on the effectiveness of the credit.[20]

The Revenue Reconciliation Act, Title VII of the Omnibus Budget Reconciliation Act of 1989, extended the tax credit through September, 1990. This law also increased the deduction disallowance to 100 percent of the research credit. Thus, the maximum effective legislated rate for the tax credit fell to 13.2 percent.[21] However, this act separated current R&E spending from the calculation of the base.

The base for the calculation of the credit was redefined to be calculated in terms of two elements. One element is the fixed base percentage which equals the ratio of qualified R&E spending to total gross receipts during the same period from 1984 through 1988. This may not exceed 16 percent. The other element is average annual gross receipts for the previous four years. The fixed base for the calculation of the R&E credit is the product of the fixed base percentage and average annual gross receipts for the past four years. As in previous legislation, the fixed base may not be less than 50 percent of qualified R&E for the current year.[22] Also, new start-up companies were included as eligible to receive this credit if their research is related to future trade or business. For such firms, the fixed base is 3 percent.[23]

The R&E tax credit incorporated in ERTA in 1981, and its subsequent revisions, cannot be viewed in isolation from the accelerated cost recovery system (ACRS) for capital expenditures first established by ERTA, but later amended by the Tax Equity and Fiscal Responsibility Act of 1982 (TEFRA).[24] On the one hand, there are at least four principal ways that changes in cost recovery can affect R&E spending. First, depreciable capital is a significant component of some types of R&E, so that these types benefit considerably from ACRS. In general, however, the capital intensity of R&E projects will differ so that the impact of changes in cost-recovery rules will be greater for some types of R&E projects than for others. Second, if depreciable assets used in R&E are treated differently from other depreciable assets, as was the case in ACRS, the required pre-tax return that must be earned by R&E relative to other investment activities must be higher. Third, because cost-recovery rules affect business cash flows and, hence, business liquidity, R&E may be affected along with other investments that are sensitive to liquidity. Finally, if changes in cost-recovery rules move the tax treatment of non-R&E business capital either closer or further from expensing, the relative tax value of R&E expensing will either be reduced or enhanced.

On the other hand, the R&E tax credit applies to the non-capital portion of R&D. Thus, it should have a proportionally greater impact on labor-intensive R&E. To some extent, at least, these two subsidies for R&E -- with their respective biases toward particular categories or types of R&E -- may each cancel the other's tendency to misallocate resources, but net biases will most likely exist.

## Effectiveness of the R&E Tax Credit

The empirical evidence on the effectiveness of the R&E tax credit is mixed. There is general agreement that R&D spending did increase in the post-1981 period, but there is disagreement as to whether this increasing trend actually began before the tax credit went into effect; or, if it was induced by the tax credit, by how much. Obviously, there is also disagreement about the net cost of the program. Because of varying opinions, and because of the likelihood of the tax credit being made permanent in the very near future, we briefly review these empirical studies.

Regarding methodologies, a number of researchers employed time series analyses in order to estimate the 1981 tax credit's initial impact on company-financed R&D spending. There are four notable studies of this type. Each relies on historical R&D-related data as the basis for formulating a forecast of post-1980 R&D spending. Then, forecasted R&D spending is compared to actual R&D spending. When actual spending exceeds the forecasted amount, it is presumed that the R&E tax credit had a positive causal impact.

Baily, Lawrence and DRI (1985) found that R&D spending was, on average, 7.3 percent greater in 1982-1983 in 12 R&D-intensive manufacturing industries than would have been predicted from historical data. Likewise, Mansfield (1985) found that actual firm-level R&D spending exceeded projected R&D spending in six R&D-intensive manufacturing industries by an average of 10 percent in 1981 and 23 percent in 1982. Brown's (1985) aggregate study reached a similar conclusion. He reported that the 1981 R&D credit was responsible for a 25 percent increase in R&D spending by 1984. Most recently, Cordes (1989) reanalyzed this issue using aggregate data and found that actual R&D spending exceeded forecasted R&D spending by 8.7 percent in 1981, 17.4 percent in 1982, 25.5 percent in 1983, and 26.8 percent in 1984. Cordes did not claim from this increasing trend that the credit was becoming more effective over time.

These four econometric studies are somewhat at odds with several related microeconomic investigations. For example, Mansfield (1985) surveyed 110 manufacturing firms. From his information it appears that the tax credit induced an increase in company-financed R&D spending by only a modest amount -- between 0.1 percent and 0.6 percent in 1981, between 0.4 percent and 1.5 percent in 1982, and between 0.6 and 1.8 percent in 1983.[25]

Eisner, Albert, and Sullivan (1983,1984), using multiple data sources, also found very little impact from the 1981 credit on R&D spending. In fact, they present rather detailed evidence that the surge in qualified R&D in 1981 was nearly offset by a decrease in non-qualified R&D. They admit, however, that this offset could either be a credit-induced substitution or a redefinition issue.[26]

Cordes (1989) attempted to reconcile the varying conclusions drawn from the time series analyses and from the microeconomic studies.[27] He notes, first, that the time series analyses may yield upwardly biased estimates if, in fact, firms are reclassifying activities as R&D. Mansfield (1985) reports from his sample of manufacturing firms that measured R&D increased by about 4 percent between 1982-1983 due to such redefining. Second, as with any time-series extrapolation, it is extremely difficult to attribute a structural change to one particular factor when many things are changing over time and across industry. Cordes acknowledges that an upward shift in aggregate R&D spending may have begun even before that 1981 credit went into effect.[28] And third, the long-run behavior response to the credit may be measured most correctly using the price elasticity of demand for R&D. And, as mentioned above, Mansfield (1985) and Baily, Lawrence, and DRI (1985) found the response, so measured, to be relatively small.

Just as there are varying opinions as to the effect of the 1981 tax credit on increasing company-financed R&D, there are varying opinions as to the cost of the tax credit. According to Mansfield (1986), his survey information from U.S., Canadian, and Swedish firms is quite similar -- the ratio of tax credit induced R&D to foregone governmental revenue is in the range of 0.3 to 0.4. Of similar magnitude, the GAO estimated that the 1981 tax credit stimulated between $1 billion and $2.5 billion of additional R&D between 1981 and 1985 at a cost of $7 billion. In other words, the credit stimulated between 15 cents and 36 cents of additional R&D spending per dollar of foregone tax revenue.[29]

In comparison, the Baily, Lawrence, and DRI's calculations include not only the direct revenue loss from the credit but also the potential revenue gains from additional taxes. They estimate that by 1991, the 1981 tax credit would result, in a worst case scenario, in a $200 million revenue loss and, in a best case scenario, in a $4.2 billion net revenue gain. Similarly, Cordes (1989) reports that he found in his 1986 National Science Foundation-sponsored study that if the 1981 tax credit were made permanent, it may have a stimulating effect of between

35 cents and 93 cents of additional R&D per each $1 of revenue loss.

Finally, the Bush Administration estimates that if the tax credit is made permanent at 20 percent beginning in 1991, revenue losses would be $0.5 billion in 1991, $0.9 billion in 1992 and $1.1 billion in 1993.[30]

## CONCLUSIONS

According to the 1981 House Report 4242, the purpose of the R&E tax credit (and presumably its renewals) was "to reverse [a] decline in research spending by industry" and "to overcome the reluctance of many ongoing companies to bear the significant costs of staffing and supplies, and certain equipment expenses such as computer charges, which must be incurred to initiate or expand research programs in trade or business."[31] It is not surprising that the Reagan Administration was concerned about declining R&D, especially when productivity growth had been declining since the mid-1960s.[32] It is also not surprising that this concern has remained during the Bush Administration given the increase in technology-based global competition.

However, efforts to increase R&D spending may not have been the correct policy target variable. First, tax incentives hold little promise for distinguishing between the total level of R&D expenditures and that portion which are successful (i.e., that actually lead to an innovation).[33] Second, not all categories of qualified R&D spending have the same measured impact on productivity growth. If one adheres to the traditional NSF data reporting categories of basic research, applied research, and development, then basic research should be targeted, at the margin, in favor of applied research or development.[34] Or, using the results presented in our earlier chapters, if productivity growth and global competitiveness are ultimate policy objectives, then perhaps investments in infratechnology should have been signaled out instead of company-financed R&D. Indeed, as noted in Chapter 1, infratechnology investments provide the bigger payoff because firms underinvest in that category of research owing to their inability to appropriate fully the benefits.

Finally, and this point applies not only as a criticism toward the R&E tax credit, in general, but also to any narrowly-focused innovation policy, R&D is one of several important inputs into a firm's innovation process.

Increasingly, technology-based firms are becoming concerned about the effective use of technical knowledge originating outside the firm. Such externally-produced technical knowledge comes from universities, state-based science and technology centers, cooperative R&D programs, joint ventures, consortia, and Federal laboratories. As we have shown in previous chapters, these external sources yield significant returns. Firms look toward an effective mix of external and internal technical knowledge as a strategic response to shortening life cycles and increasing global competition. In other words, technology-based firms strive toward developing what may be refer to as an optimal "technical information portfolio."

Theoretically, innovation policy should be aimed at increasing the efficiency of the entire technical information portfolio of firms. However, as might be expected, the administrative mechanisms with which to accomplish this are overly cumbersome. While R&D tax credits do have the potential advantage of avoiding bureaucratic entanglements, they affect portfolios differently. True, tax credits do lower the price of conducting R&D. This was noticed in the studies summarized above that relied on a price elasticity of demand for R&D in order to quantify the impact of the 1981 R&E tax credit. But, we have serious reservations about the accuracy with which these elasticities were estimated. Certainly, they cannot be determined absent a full understanding of firms' technical information portfolios -- and those portfolios probably change over time as do a firm's competitors -- and absent a complete model relating R&D to other innovative factors (as presented in Chapter 3).

## NOTES

1. This section draws from Bozeman and Link (1984, 1985).

2. See Committee for Economic Development (1980).

3. For example, formula grants to aid public education are sometimes quite complex and yet free from such controls. For an analysis of one such program and the problems of measuring its economic effects, see Leyden (1992).

4. See, for example, Surrey (1973).

5. See, for example, Cole (1971).

6. See Surrey (1969).

7. See, for example, Brimmer and Company (1979).

8. See Bozeman and Link (1983) and Link, Tassey, and Zmud (1983) for a more detailed discussion of this so-called make versus buy issue.

9. The National Science Foundation (NSF) data reporting definition of R&D is basic and applied research in the sciences and engineering and the design and development of prototypes and processes. This definition excludes quality control, routine product testing, market research, sales promotion, sales service, research in the social sciences or psychology, and other nontechnological activities or routine technical services. Based on the IRS regulations, R&E is defined as expenditures which represent NSF-defined research and development costs in the experimental or laboratory sense. The term R&E generally includes all costs incident to the development of an experimental or pilot model, a plant process, a product, formula, and invention or a similar property. The term is not intended to refer to ordinary testing or inspection of materials or products for quality control or those for efficiency surveys, management studies, consumer surveys, advertising, or promotions (KPMG, 1990).

10. The term depreciation refers to the accounting process of systematically allocating the cost of a long-lived asset over its useful life.

11. See, for example, Kaplan (1975).

12. See Kaplan (1975).

13. Since 1977, Treasury Regulation 1.861-8 has required U.S. multinational firms to allocate some of their domestic R&E expenditures against income from foreign sources. The rationale is, if a firm spends money for R&E in the U.S. and the resulting products or processes are sold abroad, then a portion of these R&E costs should be allocated against foreign sales. The combined effect of this regulation and the tax laws governing foreign income is to increase the effective tax rate on foreign income and perhaps to encourage multinational firms to export a portion of their R&E overseas. See Ruscio (1981).

14. See Eisner, Albert, and Sullivan (1984).

15. According to Baily and Lawrence (1990) this credit reduced, on an after-tax basis, the cost of qualified R&E by 9.3 percent.

16. See U.S. General Accounting Office (1989). The term company-financed R&D refers to those R&D expenditures financed internally, as opposed to those being financed through governmental contracts or grants.

17. The remainder of this section draws from Hankins and Scheirer (1989), Wozny (1989), U.S. General Accounting Office (1989) and U.S. Department of Commerce, Office of Technology Policy (1990). According to Baily and Lawrence (1990), the rate reduction to 20 percent coupled with a lowering of corporate income tax rates implied that the credit, on an after-tax basis, reduced the cost of qualified R&E from 9.3 percent to 6.7 percent in 1987 and to 6.1 percent in 1988.

18. The National Advisory Committee on Semiconductors has recommended that the tax credit be extended to cover commercial R&D conducted in consortia, such as SEMATECH. See Congressional Budget Office (1990).

19. Again, according to Baily and Lawrence (1990) the after-tax cost reduction fell to 4 percent.

20. See U.S. General Accounting Office (1989).

21. See KPMG (1990) for examples.

22. For examples see KPMG (1990).

23. For a discussion of the importance of making the credit applicable to start-up companies, see Swain (1988).

24. An empirical analysis of the relative incentives established by each of these aspects of ERTA is in Link and Tassey (1987). See also Barth, Cordes, and Tassey (1984).

25. Mansfield notes that these estimates may approximate the expected long-run impact from the credit. A reasonable estimate of the effective tax credit is 6 percent. Eisner, Albert, and Sullivan (1983) and U.S. General Accounting Office (1989) estimate the effective rate to be closer to 4 percent. If the price elasticity of the demand for R&D is 0.3, as suggested by those studies so referenced, then a 1.8 percent annual increase in R&D appears to be reasonable. Mansfield (1986) also notes in a related review article that others have found that the credit had only a modest impact on R&D spending of between 0.4 percent and 0.8 percent per year. See, Charles River Associates (1985).

26. See Altshuler (1988) for a discussion of the calculation of effective rates under alternative financial scenarios.

27. See Gravelle (1985) and Collins (1986), too.

28. See Papadakis (1990) for empirical support. In fact, her analyses show that the upward shift began in the 1977-1978 period.

29. See U.S. Department of Commerce, Office of Technology Policy (1990).

30. See U.S. General Accounting Office (1989).

31. Quoted from U.S. Department of Commerce, Office of Technology Policy (1990).

32. See Link (1987) for a review of this literature.

33. We know that firm size is not a prerequisite for success in R&D beyond a modest threshold level (Link, 1980), and we know that larger firms, generally those with corporate assets greater than $250 million, took greater advantage of the R&D tax credit than did smaller firms (Eisner, Albert, and Sullivan, 1984; U.S. General Accounting Office, 1989).

34. Mansfield (1980) and Link (1981) have verified this correlation. Also, Bozeman and Link (1984, 1985) proposed a tax credit for cooperative research investments (cooperative research occurs at the basic end of the R&D spectrum). Link and Bauer's (1989) econometric analyses demonstrate the productivity growth increases associated with cooperative research endeavors.

# 8
# Restating Government's Role in Innovation

The Administration's formal technology policy, as stated in the Executive Office's *U.S. Technology Policy*, emphasized the government's role in three key areas, although remaining somewhat vacuous in terms of the specifics for an implementation strategy. The key areas that are emphasized, and rightly so in our estimation, are:

- funding of R&D performed in the private sector;
- funding of Federal laboratory research activities and the effective transfer of that technology to the private sector; and
- funding of basic research (especially at universities) and encouragement of industry-university research relationships.

On the basis of a theoretical and empirical investigation of the overall innovation process (Chapter 3), an econometric examination of the interactions between privately-financed and Federally-financed R&D (Chapters 3, 4, and 6), and case analyses of the economic impacts of infratechnology research (Chapter 5), we conclude that the government's continued efforts in the three areas noted in the Administration's technology policy are warranted. The reason that these governmental efforts have been so successful is that they all relate to the building of our Nation's technology infrastructure through investment activities that emphasize infratechnology research. This is important because infratechnology research facilitates the entire R&D production process, and because it is the one type of innovation-related activity that private-sector firms are least willing to undertake alone.

Although the permanency of the R&D tax credit will be viewed by many as a major technology policy step, we are skeptical of its effectiveness. As discussed in Chapter 7, there are pros and cons associated with any tax incentive. The existing empirical evidence of the effectiveness of the tax credit in the past has been mixed,[1] and strong arguments can be made that the R&D tax credit is indiscriminate in its effects on private-sector R&D activity.

We argue, instead, that future attention should focus not on R&D-specific indirect policies, but rather on policies that have the ability to build the Nation's technology infrastructure. One such existing program is the Department of Commerce's Advanced Technology Program.

The Omnibus Trade and Competitiveness Act of 1988, P.L. 100-418 (hereafter referred to as the Act), established the Advanced Technology Program (ATP). See Section 28(a). This part of the Omnibus Trade and Competitiveness Act of 1988 is also known as the Technology Competitiveness Act.

As stated in the Act, the goals of the ATP are to assist U.S. businesses in creating and applying the generic technology and research results necessary to:

- commercialize significant new scientific discoveries and technologies rapidly; and

- refine manufacturing technologies.

These goals were also restated in the *Federal Register* on July 24, 1990:

> The ATP ... will assist U.S. businesses to improve their competitive position and promote U.S. economic growth by accelerating the development of a variety of pre-competitive generic technologies by means of grants and cooperative agreements.

To accomplish these goals, the ATP solicited and evaluated proposals (proposal solicitation 90-01) under the guidelines stated in the *Federal Register* on July 24, 1990. On March 5, 1991, the ATP announced its intention to fund projects submitted by 11 of the nearly 250 applicants. Six awards were given to single businesses and five to cooperative research ventures. Total Federal funding for these projects is estimated to be $46 million over 5 years, with an additional $52 million in "cost-sharing" provided by the private sector. Project managers from the ATP staff will monitor the progress of the funded research. The second solicitation announcement was published on June 28, 1991.

In our opinion this policy initiate is correctly designed. It not only emphasizes infratechnology research as the primary driver of technology-based growth, but also it identifies cooperation as the proper mechanism to conduct the research.[2] A sound technology infrastructure is the single most important ingredient in positioning an industrial nation to meet the global challenges in the decades ahead.

## NOTES

1. See Bozeman and Link (1983, 1985) for an alternative tax credit-based policy suggestion. Basically, they argue for a tax credit for undertaking cooperative research.

2. See Link and Bauer (1989).

# References

Altshuler, R., "A Dynamic Analysis of the Research and Experimentation Credit," *National Tax Journal* (1988): 453-456.

Augustine, N.A., *Report of the Defense Science Board Task Force on Semiconductor Dependency*, Department of Defense, Washington, D.C., 1987.

Baily, M.N. and R.Z. Lawrence, "The Incentive Effects of the New R&E Tax Credit," mimeographed, 1990.

Baily, M.N., R.Z. Lawrence, and Data Resources, Inc. (DRI), "The Need for a Permanent Tax Credit for Industrial Research and Development," contract report prepared for the Coalition for the Advancement of Industrial Technology, 1985.

Barth, J., J.J. Cordes, and G. Tassey, "The Impact of Recent Changes in Tax Policy on Innovation and R&D" in *Strategic Management of R&D*, edited by B. Bozeman, M. Crow, and A.N. Link, Lexington Books, Lexington, Mass., 1984.

Belsley, D.A., "Two- or Three-Stage Least Squares," *Computer Science in Economics and Management* (1988): 21-30.

Bendor, J., S. Taylor, and R. van Gaalen, "Bureaucratic Expertise Versus Legislative Authority: A Model of Deception and Monitoring in Budgeting," *American Political Science Review* (1985): 1041-1060.

Blank, D.M. and G.J. Stigler, *The Demand and Supply of Scientific Personnel*, National Bureau of Economic Research, New York, 1957.

Bozeman, B. and M, Crow, "U.S. R&D Laboratories and Their Environments: Public and Market Influence," final report to the National Science Foundation, 1988.

Bozeman, B. and A.N. Link, *Investments in Technology: Corporate Strategies and Public Policy Alternatives*, Praeger, New York, 1983.

Bozeman, B. and A.N. Link, "Public Support for Private R&D: The Case of the Research Tax Credit," *Journal of Policy Analysis and Management* (1985): 370-382.

Bozeman, B. and A.N. Link, "Tax Incentives for R&D: A Critical Evaluation," *Research Policy* (1984): 21-31.

Brimmer and Company, Inc., "Technology, Innovation, and Economic Growth," mimeographed, 1979.

Brown, K.M., *The R&D Tax Credit: An Evaluation of Evidence on Its Effectiveness*, Joint Economic Committee, Congress of the United States, Washington, D.C., 1985.

Burness, H.S., W.D. Montgomery, and J.P. Quirk, "Capital Contracting and the Regulated Firm," *American Economic Review* (1980): 342-354.

Carlson, E., "Government Work Can Be a Boon, But it Isn't Risk-Free: Producer for TV Marti Learns Lessons About Relying on One U.S. Contract," *The Wall Street Journal* (May 16, 1991): 32.

Carmichael, J., "The Effects of Mission-Oriented Public R&D Spending on Private Industry," *Journal of Finance* (1981): 617-627.

Carnegie Commission on Science, Technology, and Government, *Technology and Economic Performance: Organizing the Executive Branch for a Stronger National Technology Base*, Carnegie Commission, New York, 1991.

Chaffee, C.D., *The Rewiring of America: The Fiber Optics Revolution*, Academic Press, Orlando, Fla., 1988.

Charles River Associates, "An Assessment of Options for Restructuring the R&D Tax Credit to Reduce Dilution of its Marginal Incentive," mimeographed, 1985.

Cohen, L.R. and R.G. Noll, *The Technology Pork Barrel*, The Brookings Institution, Washington, D.C., 1991.

Cohen, W.M. and D.A. Levinthal, "The Implication of Spillovers for R&D Investment and Welfare" in *Advances in Applied Microeconomics*, edited by A.N. Link and V.K. Smith, JAI Press, Greenwich, Conn., 1990.

Cole, R.T., "Discussion" in *Tax Incentives*, edited by S.S. Surrey, Lexington Books, Lexington, Mass., 1971.

Collins, E.L., "Sorting Out the Economic Arguments Underlying Proposed Tax Incentives to Encourage Innovation," mimeographed, 1980.

Collins, E.L., "Tax Policy and Industrial Innovation: America's Experience with an R&D Tax Credit" in *Technological Innovation: Strategies for a New Partnership*, edited by D.O. Gray, T. Solomon, and W. Hetzner, North Holland, Amsterdam, 1986.

Committee for Economic Development, *Stimulating Technological Progress*, Committee for Economic Development, Washington, D.C., 1980.

Congressional Budget Office, *Using R&D Consortia for Commercial Innovation*, Government Printing Office, Washington, D.C., 1990.

Cordes, J.J., "Tax Incentives and R&D Spending: A Review of the Evidence," *Research Policy* (1989): 119-133.

Council on Competitiveness, "Council Poll: Voters Ready For 'Straight Talk' on Competitiveness," *Challenges* (December 1991): 1-6.

Council on Competitiveness, *Picking Up the Pace: The Commercial Challenge to American Innovation*, Washington, D.C., 1989.

Dasgupta, P., "The Economic Theory of Technology Policy: An Introduction" in *Economic Policy and Technology Performance*, edited by P. Dasgupta and P. Stoneman, Cambridge University Press, Cambridge, 1987.

Eisner, R., S.H. Albert, and M.A. Sullivan, "Tax Incentives for R&D Expenditures," mimeographed, 1983.

Eisner, R., S.H. Albert, and M.A. Sullivan, "The New Incremental Tax Credit for R&D: Incentive or Disincentive?" *National Tax Journal* (1984): 171-183.

Engineering Research Board, *Directions in Engineering Research -- An Assessment of Opportunities and Needs*, National Academy Press, Washington, D.C., 1987.

Executive Office of the President, *U.S. Technology Policy*, Office of Science and Technology Policy, Washington, D.C., September 26, 1990.

Ferguson, C.E., *The Neoclassical Theory of Production and Distribution*, Cambridge University Press, Cambridge, 1979.

Gomez-Ibanez, J.A., J.R. Meyer, and D.E. Luberoff, *The Prospects for Privatizing Infrastructure: Lessons from U.S. Roads and Solid Waste*, mimeographed, 1990.

Gravelle, J.G., "The Tax Credit for Research and Development: An Analysis," mimeographed, 1985.

Griliches, Z., "Issues in Assessing the Contribution of Research and Development to Productivity Growth," *Bell Journal of Economics* (1979): 92-116.

Griliches, Z., "Productivity Puzzles and R&D: Another Nonexplanation," *Journal of Economic Perspectives* (1988): 9-22.

Hankins, M.D. and W.K. Scheirer, "Innovation, Government, and Small Business," mimeographed, 1989.

Hicks, J.R., *Value and Capital: An Inquiry into Some Fundamental Principles of Economic Theory*, second edition, Oxford University Press, Oxford, 1948.

International Trade Administration (ITA), *A Competitive Assessment (Update) of the U.S. Fiber Optics Industry*, Washington, D.C., 1988.

International Trade Commission (ITC), *U.S. Global Competitiveness: Optical Fibers, Technology and Equipment*, Washington, D.C., 1988.

Joskow, P.L., "Inflation and Environmental Concern: Structural Change in the Process of Public Utility Price Regulation," *Journal of Law and Economics* (1974): 291-327.

Kaplan, R.S., *Tax Policies for R&D and Technological Innovation*, National Technical Information Services, Washington, D.C., 1975.

Klynveld Peat Marwick Goerdeler (KPMG), *Tax Treatment of Research and Development Expenditures*, International Tax Center, Amsterdam, 1990.

Krasa, E.M., "Research and Development Tax Changes: New Opportunities for Performers and Investors," *Business Quarterly* (1983): 35-38.

Levin, R.C., W.M. Cohen, and D.C. Mowery, "R&D Appropriability, Opportunity, and Market Structure: New Evidence on Some Schumpeterian Hypotheses," *American Economic Review* (1985): 20-24.

Levin, R.C., A.K. Klevorick, R.R. Nelson, and S.G. Winter, "Appropriating the Returns from Industrial Research and Development," *Brookings Papers on Economic Activity* (1987): 783-820.

Levin, R.C. and P.C. Reiss, "Tests of a Schumpeterian Model of R&D and Market Structure" in *R&D, Patents, and Productivity*, edited by Z. Griliches, University of Chicago Press, Chicago, 1984.

Levy, D.M., "Estimating the Impact of Government R&D," *Economics Letters* (1990): 169-173.

Levy, D.M. and N.E. Terleckyj, "Effects of Government R&D on Private R&D Investments and Productivity: A Macroeconomic Analysis," *Bell Journal of Economics* (1983): 551-561.

Leyden, D.P., "Linear-Demand Models of Intergovernmental Grants Versus a More Structured Approach: An Empirical Evaluation," mimeographed, 1992.

Leyden, D.P. and A.N. Link, "Privatization, Bureaucracy, and Risk Aversion," *Public Choice* (1992): forthcoming.

Leyden, D.P. and A.N. Link, "Why are Governmental R&D and Private R&D Complements?" *Applied Economics* (1991): 1673-1681.

Leyden, D.P., A.N. Link, and B. Bozeman, "The Effects of Governmental Financing on Firms' R&D Activities: A Theoretical and Empirical Investigation," *Technovation* (1989): 561-575.

Lichtenberg, F.R., "The Effect of Government Funding on Private Industrial Research and Development: A Re-Assessment," *Journal of Industrial Economics* (1987): 97-104.

Lichtenberg, F.R., "The Relationship Between Federal Contract R&D and Company R&D, *American Economic Review* (1984): 73-78.

Lindsay, C.M., "A Theory of Government Enterprise," *Journal of Political Economy* (1976): 1061-1077.

Link, A.N., "An Analysis of the Composition of R&D Spending," *Southern Economic Journal* (1982): 342-349.

Link, A.N., "Basic Research and Productivity Increase in Manufacturing: Some Additional Evidence," *American Economic Review* (1981): 1111-1112.

Link, A.N., "Firm Size and Efficient Entrepreneurial Activity: A Reformulation of the Schumpeter Hypothesis," *Journal of Political Economy* (1980): 771-782.

Link, A.N., "Investments in Infratechnology as an Indicator of Technological Advancement: An Exploratory Study," final report to the National Science Foundation Science and Engineering Indicators Program, November 1991.

Link, A.N., *Technological Change and Productivity Growth*, Harwood Academic Publishers, London, 1987.

Link, A.N. and L.L. Bauer, *Cooperative Research in U.S. Manufacturing: Assessing Policy Initiatives and Corporate Strategies*, Lexington Books, Lexington, Mass., 1989.

Link, A.N., B. Bozeman, and D.P. Leyden, "Federal R&D and Industrial Innovative Activity," final report to the National Science Foundation Economic Analysis Studies Group, 1990.

Link, A.N. and J.E. Long, "The Simple Economics of Basic Scientific Research: A Test of Nelson's Diversification Hypothesis," *Journal of Industrial Organization* (1981): 105-109.

Link, A.N., P. Quick, and G. Tassey, "Investments in Product Quality: A Descriptive Study of the U.S. Optical Fiber Industry," *Omega: The International Journal of Management Science* (1991): 471-474.

Link, A.N. and J. Rees, "Firm Size, University Based Research, and Returns to R&D," *Small Business Economics* (1990): 25-31.

Link, A.N., T.G. Seaks, and S. Woodbery, "Firm Size and R&D Spending," *Southern Economic Journal* (1988): 1027-1032.

Link, A.N. and G. Tassey, *Strategies for Technology-based Competition: Meeting the New Global Challenge*, Lexington Book, Lexington, Mass., 1987.

Link, A.N., G. Tassey, and R. Zmud, "The Induce Versus Purchase Decision: An Empirical Analysis of Industrial R&D," *Decision Sciences* (1983): 46-61.

McCubbins, M.D. and T. Page, "A Theory of Congressional Delegation" in *Congress: Structure and Policy*, edited by M.D. McCubbins and T. Sullivan, Cambridge University Press, Cambridge, 1987.

McElroy, M., "Goodness of Fit for Seemingly Unrelated Regressions: Glahn's $R^2_{yx}$ and Hooper's $r^2$," *Journal of Econometrics* (1977): 381-387.

Magaziner, I. and M. Patinkin, *The Silent War*, Random House, New York, 1989.

Mansfield, E., "Basic Research and Productivity Increase in Manufacturing," *American Economic Review* (1980): 863-873.

Mansfield, E., "Public Policy Toward Industrial Innovation: An International Study of Direct Tax Incentives for Research and Development" in *The Uneasy Alliance: Managing the Productivity-Technology Dilemma*, edited by K.B. Clark, R.H. Hayes, and C. Lorenz, Harvard Business School Press, Cambridge, 1985.

Mansfield, E., "The R&D Tax Credit and Other Technology Policy Issues," *Research Policy* (1986): 190-194.

Mansfield, E. and L. Switzer, "Effects of Federal Support on Company Financed R&D: The Case of Energy," *Journal of Management Science* (1984): 562-571.

Mansfield, E. and L. Switzer, "The Effects of R&D Tax Credits and Allowances in Canada," *Research Policy* (1985): 97-105.

Mueller, D.C., *Public Choice II*, Cambridge University Press, Cambridge, 1989.

Musgrave, R.A. and P.B. Musgrave, *Public Finance in Theory and Practice*, fifth edition, McGraw-Hill, New York, 1989.

National Science Board, *Science & Engineering Indicators - 1991*, Government Printing Office, Washington, D.C., 1991.

National Science Foundation, *Research and Development in Industry: 1988*, National Science Foundation, Washington, D.C., 1990.

Nelson, R.R., "The Simple Economics of Basic Scientific Research," *Journal of Political Economy* (1959): 297-306.

Niskanen, W.A., Jr., *Bureaucracy and Representative Government*, Aldine-Atherton, Chicago, 1971.

Papadakis, M., "Bringing Science to Market: the Policy Implications of U.S. and Japanese Patterns of Science, Technology, and Competitiveness," mimeographed, 1990.

Quick, Finan & Associates, "U.S. Investment Strategies for Quality Assurance," NIST Planning Report 90-1, 1990.

Rasmussen, E., *Games and Information: An Introduction to Game Theory*, Basil Blackwell, Oxford, 1989.

Raviv, A., "The Design of an Optimal Insurance Policy," *American Economic Review* (1979): 84-96.

Rees, J. (ed.), *Technology, Regions and Policy*, Rowman and Littlefield, Totowa, N.J., 1986.

Ross, I.M., *A Strategic Industry at Risk*, A Report to the President and Congress from the National Committee on Semiconductors, Washington, D.C., 1989.

Ruscio, K., "Tax Incentives and Innovation," mimeographed, 1981.

Sappington, D.E.M., "Incentives in Principal-Agent Relationships," *Journal of Economic Perspectives* (1991): 45-66.

Sappington, D.E.M. and J.E. Stiglitz, "Privatization, Information and Incentives," *Journal of Policy Analysis and Management* (1987): 567-582.

Savas, E.S., *Privatization: The Key to Better Government*, Chatham House, Chatham, N.J., 1987.

Scace, R.I., "Metrology for the Semiconductor Industry," mimeographed, undated.

Scace, R.I., "Semiconductors," mimeographed, 1990.

Schafft, H.A., "CEEE Success Story," mimeographed, undated.

Schafft, H.A., "Examples of the Impact of Our Work to the Semiconductor Community," mimeographed, 1989.

Schafft, H.A., "Response to Request for Impact Statements for Use in Testimony," mimeographed, 1985.

Schafft, H.A., "Thermal Analysis of Electromigration Test Structures," *IEEE Transactions on Electron Devices* (1987): 664-672.

Schafft, H.A. and J. Albers, "Thermal Interactions Between Electromigration Test Structures," Proceedings of the 1988 IEEE International Conference on Microelectronic Test Structures, 1988.

Schafft, H.A., J. Lechner, B. Sabi, M. Mahaney, and R. Smith, "Statistics for Electromigration Testing," Proceedings of the International Reliability Physics Symposium, 1988.

Schafft, H.A., T.C. Staton, J. Mandel, and J.D. Shott, "Reproducibility of Electromigration Measurements," *IEEE Transactions on Electron Devices* (1987): 673-681.

Scott, J.T., "Firm versus Industry Variability in R&D Intensity" in *R&D, Patents, and Productivity*, edited by Z. Griliches, University of Chicago Press, Chicago, 1984.

Senior, J., *Optical Fiber Communications: Principles and Practices*, Prentice-Hall, London, 1985.

Simon, H.A., "Organizations and Markets," *Journal of Economic Perspectives* (1991): 25-44.

Stiglitz, J.E., *Economics of the Public Sector*, second edition, W.W. Norton, New York, 1988.

Stiglitz, J.E., "Symposium on Organizations and Economics," *Journal of Economic Perspectives* (1991): 15-24.

Suehle, J.S. and H.A. Schafft, "Current Density Dependence of Electromigration $t_{50}$ Enhancement Due to Pulsed Operations," Proceedings of the International Reliability Physics Symposium, 1990.

Suehle, J.S. and H.A. Schafft, "The Electromigration Damage Response Time and Implications for DC Pulsed Characterizations," Proceedings of the International Reliability Physics Symposium, 1989.

Surrey, S.S., *Pathways to Tax Reform: The Concept of Tax Expenditures*, Harvard University Press, Cambridge, 1973.

Surrey, S.S., "Tax Incentives: Conceptual Criteria for Identification and Comparison with Direct Government Expenditures," *Proceedings of the Tax Institute of America*, 1969.

Swain, F.S., "Testimony before the Senate Subcommittee on Taxation and Debt Management," mimeographed, 1988.

Tassey, G., "The Functions of Technology Infrastructure in a Competitive Economy," *Research Policy* (1991): 345-361.

Tassey, G., "Semiconductors," mimeographed, undated.

Tassey, G., "Structural Change and Competitiveness: The U.S. Semiconductor Industry," *Technological Forecasting and Social Change* (1990): 85-93.

Tassey, G., *Technology Infrastructure and Competitive Position*, Kluwer Academic Publishers, Norwell, Mass., 1992.

Terleckyj, N.E., *Effects of R&D on the Productivity Growth of Industries: An Exploratory Study*, National Planning Association, Washington, D.C., 1974.

Tullock, G., "Rent-Seeking as a Negative-Sum Game" in *Toward a Theory of the Rent-Seeking Society*, edited by J.M. Buchanan, R.D. Tollison, and G. Tullock, Texas A&M Press, College Station, Tx., 1980.

U.S. Department of Commerce, *The Competitive Status of the U.S. Electronics Sector from Materials to Systems*, Report from the Secretary of Commerce to the Appropriations Committee of the U.S. House of Representatives, Washington, D.C., April 1990.

U.S. Department of Commerce, *Emerging Technologies: A Survey of Technical and Economic Opportunities*, Government Printing Office, Washington, D.C., 1990.

U.S. Department of Commerce, Office of Technology Policy," "Analysis of the Research Tax Credit," mimeographed, 1990.

U.S. General Accounting Office, *Tax Policy and Administration: The Research Tax Credit Has Stimulated Some Additional Research Spending*, Government Printing Office, Washington, D.C., 1989.

U.S. Small Business Administration, Office of Advocacy, *Innovation in Small Firms*, Washington, D.C., 1986.

Varian, H.R., *Microeconomic Analysis*, second edition, W.W. Norton, New York, 1984.

Weiss, L.W. and G.A. Pascoe, "Adjusted Concentration Ratios in Manufacturing, 1972 and 1977," mimeographed, 1986.

Weingast, B.R., "The Congressional-Bureaucratic System: A Principal-Agent Perspective (with Application to the SEC), *Public Choice* (1984): 147-191.

Wilson, J.Q., *Bureaucracy: What Government Agencies Do and Why They Do It*, Basic Books, New York, 1989.

Wozny, J.A., "The R&D Tax Credit: An Evaluation of Recent Revisions and Proposals," mimeographed, 1989.

# Index

- F -



- G -



- H -



- I -



- J -



- K -



- L -

- M -



- N -



- O -



- P -

**- Q -**



**- R -**



**- S -**

- T -



- U -



- V -



- W -



- X -



- Z -